The Container Principle

Infrastructures Series

edited by Geoffrey Bowker and Paul N. Edwards

Paul N. Edwards, *A Vast Machine: Computer Models, Climate Data, and the Politics of Global Warming*

Lawrence M. Busch, *Standards: Recipes for Reality*

Lisa Gitelman, ed., *"Raw Data" Is an Oxymoron*

Finn Brunton, *Spam: A Shadow History of the Internet*

Nil Disco and Eda Kranakis, eds., *Cosmopolitan Commons: Sharing Resources and Risks across Borders* Casper Bruun Jensen and Brit Ross Winthereik, *Monitoring Movements in Development Aid: Recursive Partnerships and Infrastructures*

James Leach and Lee Wilson, eds., *Subversion, Conversion, Development: Cross-Cultural Knowledge Exchange and the Politics of Design*

Olga Kuchinskaya, *The Politics of Invisibility: Public Knowledge about Radiation Health Effects after Chernobyl*

Ashley Carse, *Beyond the Big Ditch: Politics, Ecology, and Infrastructure at the Panama Canal*

Alexander Klose, translated by Charles Marcrum II, *The Container Principle: How a Box Changes the Way We Think*

The Container Principle

How a Box Changes the Way We Think

Alexander Klose

Translated by Charles Marcrum II

The MIT Press
Cambridge, Massachusetts
London, England

Copyright © 2009 mareverlag, Hamburg

English translation © 2015 Massachusetts Institute of Technology

The translation of this work was funded by Geisteswissenschaften International–Translation Funding for Humanities and Social Sciences from Germany, a joint initiative of the Fritz Thyssen Foundation, the German Federal Foreign Office, the collecting society VG WORT, and the Börsenverein des Deutschen Buchhandels (German Publishers & Booksellers Association).

All rights reserved. No part of this book may be reproduced in any form by any electronic or mechanical means (including photocopying, recording, or information storage and retrieval) without permission in writing from the publisher.

MIT Press books may be purchased at special quantity discounts for business or sales promotional use. For information, please email special_sales@mitpress.mit.edu.

This book was set in Stone by the MIT Press. Printed and bound in the United States of America.

Library of Congress Cataloging-in-Publication Data

Klose, Alexander, 1969–
[Container-Prinzip. English]
The container principle : how a box changes the way we think / by Alexander Klose ; translated by Charles Marcrum II.
pages   cm. — (Infrastructures series)
Translation of: Das Container-Prinzip : wie eine Box unser Denken verändert. Published: Hamburg : Mare Verlag, 2009.
Includes bibliographical references and index.
ISBN 978-0-262-02857-8 (hardcover : alk. paper)
1. Containerization—History. 2. Containerization—Social aspects. 3. Container ships—History. 4. Unitized cargo systems—History. I. Title.
TA1215.K5813   2009
688.801'9—dc23
2014016713

10 9 8 7 6 5 4 3 2 1

Contents

Preface to the English Edition vii
Acknowledgments xiii

Introduction: The Accident 1
1 **Time Capsules** 9
2 **What Is a Container?** 39
3 **Sea-Land** 83
4 **Container Histories** 119
5 **Logistics: The Power of a Third Party** 155
6 **Computing with Containers** 199
7 **Life in Cells** 249
8 **Container World** 295

Notes 343
Bibliography 357
Index 377

Preface to the English Edition

Washington D.C., August 2007, post-9/11 America: The U.S. Senate passes Public Law 110–53 to enhance the security of international container shipping and to protect the American homeland. Associated Press journalist Jim Abrams, paraphrasing the famous opening line of *The Communist Manifesto*, said it was the "specter of a nuclear bomb, hidden in a cargo container" that "prompted Congress to require 100 percent screening of U.S. bound ships at their more than 600 foreign starting points."[1]

The new screening law is part of the Container Security Initiative (CSI), a subdivision of U.S. Customs and Border Protection. CSI was founded in early 2002 during the high pitch of federal homeland security activism. The official CSI website's argument for the necessity of drastic security measures for cargo transport reads as follows: "Imagine if a weapon of mass destruction sitting in a container within the sea cargo environment were detonated."[2]

After the collapse of the Berlin Wall and the destruction of the twin towers of the World Trade Center, it is no longer the specter of communism—at least for the time being—that haunts the minds of U.S. security agencies. The "imagination of disaster"[3] focuses on the driving force of U.S. economic success and supremacy. The security of the (Western) world, as the image of smuggled weapons of mass destruction suggests, is threatened by precisely the ubiquitous transport medium that enabled the economic growth and welfare of the last decades. These very same boxes were sent into worldwide circulation for a long period, unhindered as much as possible by trade barriers and complicated bureaucratic procedures, thus establishing today's global production and supply chains. Under the new regulations, they would have to be controlled and inspected again.

Germany, spring 2009, financial crisis: The front page of the May 20, 2009, issue of *Die Zeit*, one of the leading German newspapers, shows a reproduction of the digitally altered painting *Tower of Babel* by Pieter Brueghel the Elder. The headline reads "Will It Work without Growth?" The tower is being reconstructed; it is now perforated and encircled by railroad tracks and highways transporting an unending chain of containers. The standard steel stacked boxes drive the tower higher into the sky. Babel, the biblical city on many waters, is located by the sea. Container ships can be seen coming across the vast ocean. Supplies for the tower are piled up on the wharf. As this picture suggests, the hubris of modern global society is built on containers. And the way out of the permanent crisis of capitalism lies in ending the race for the most boxes.

What the two scenarios described above show is that the former black boxes of globalization have turned into thoroughly ambivalent objects. Pandora's steel boxes are not only carrying the world's cargo trade but are also depicting everything that is bad or problematic about globalization. During the last decade or two, transport containers have become critical objects; they have become visible, an opportune revelation for researchers and media archaeologists of the present time. Infrastructures, when working smoothly, tend to vanish under the threshold of visibility.

Much more urgent today than the problem of confused tongues traditionally associated with the story of the Tower of Babel is the problem of diversified and interwoven circumstances of production negotiated under the term *globalization*. The wealth of nations and their employment opportunities are significantly dependent on the continued prosperity of such global supply chains. This is evident in the performance of the core element of logistics: the container transport system and its microeconomic logic that has gripped the whole world.

Containers are not simply the most, if not the only, important means of transport for the vast majority of the goods we deal with every day. Perhaps because of their simple, clear, expressive power, containers have become the symbol of globalization—and also of many phenomena associated with this development. Containers represent the impressive dynamics of modern capitalism and its fundamental optimism in the face of every crisis. At the same time, they represent the fears of and objections to these dynamics when logistics are organized purely for optimization, forcibly converging and aligning formerly remote parts of the world through an exponential increase in transport and communication processes.

The basic material quality of containers, the fact that they can be emptied just as easily as they can be filled, also seems to reveal an effect on the semantic level of stories and images. Thus, one finds containers not only in the business sections of newspapers and television broadcasts but also in films, plays, and novels. Because of their versatility as modular spaces and because they can be assembled and disassembled comparatively easily, containers have also been nearly ubiquitous in residential areas for many years: as temporary accommodations and storage for people and materials, as business or office spaces, and as kindergartens or kiosks.

The omnipresence of containers and the surprising fact that I found very little that attempted to explain this omnipresence was my starting point as I made the voyage from Hamburg to Hong Kong in the summer of 2001 as a member of a film team. I approached the journey on the container ship as the water-travel portion of a kind of Grand Tour of globalization, to get a sense for myself of how containers affect the global economy.

At the same time, however, I was almost more interested in the containers ashore—especially since I already viewed the container ship itself as essentially a huge floating parking lot. All the container-shaped things that I encountered in the most diverse fields of society were constantly spreading, as it seemed to me—from physical storage systems to spatial organizational metaphors—and I became more and more convinced that containerization is more than the transformation of freight traffic to shipment in standard containers. What it is, in fact, is a grand movement comparable to mechanization in the breadth of its applications—a change in the fundamental order of thinking and things that may be spoken of as a principle, the material core of which is the standardized container, by which it became fully visible but in which it hardly exhausts itself.

I dedicated more than five years of research to the emergence of this principle, its prehistory, and its spread. The results of my visits to the various "container worlds"—to trade shows and harbors, to the offices of logistics experts and architects, in assorted scientific fields, in the fine arts, in films and novels, and in philosophy—form the content of this book. In light of the fact that my research was based at a German university, but also as a reaction to my finding that most publications tell the history and principles of containerization almost exclusively from the American perspective, this book often emphasizes the European (and particularly the German) side of the historical developments; this may be a bit surprising, but I hope it is also rewarding to the American reader.

If my theory of the ubiquity of the container principle is right, then it is impossible to treat the subject definitively at the present time. At best, one can propose theories and attempt to trace tendencies. Thus, the chapters of this book concentrate on "container situations" associated with certain historical and systemic conditions, such as the history of logistics or the idea of the standardized modular compartment. Each chapter undertakes a journey around the world and into the depths of history. Reflecting its particular subject, each chapter can also be read as a separate unit. The book contains eight—by no means standardized—reading modules (chapters), and it is not necessary to read them in sequence.

Acknowledgments

This book could never have come about without the help of numerous people from a variety of fields. Above all, I would like to thank Barbara Klose-Ullmann for her generous support on many levels. Furthermore, my friends and colleagues associated with the graduate research group Media Historiographies from the Universities of Weimar, Erfurt, and Jena and the Bauhaus University, foremost among them Bernhard Siegert, the advisor for my research project, Jan Philip Müller, and Anne Fleckstein, as well as the following (in alphabetical order): Alessandro Barberi, Jan Behnstedt, Thorsten Bothe, Lorenz Engell, Daniel Eschkötter, Rupert Gaderer, Moritz Gleich, Stephan Gregory, Marion Herz, Christina Hünsche, Gregor Kanitz, Isabel Kranz, Alf Lüdtke, Helga Lutz, Bettine Menke, Jakob Racek, Christoph Rosol, Ulfert Tschirner, Christina Vagt, Sebastian Vehlken, Joseph Vogl, Karoline Weber, and Nina Wiedemeyer.

For conversations, advice, and inspiration, I would like to thank the following people in particular (in alphabetical order): Andreas Bernard (Berlin and Munich), Peter Berz (Berlin), Sigi Bouvier (Munich), Lars Denicke (Berlin), Arthur Donovan (New York), Axel Dossmann (Berlin and Jena), Suse Drost (Tel Aviv), Susanne Dunke (Berlin), Klaus Ebeling (Neuilly sur

Seine and Brussels), Knut Ebeling (Berlin), Paul Edwards (Ann Arbor), Tineke M. Egyedi (Amsterdam and Delft), Martina Endres (Berlin), Alexandra Engel (Berlin), Christoph Engemann (Lüneburg, Berlin), etoy.ZAI (Zurich), Michael Freitag (Bremen), Maik Freudenberg (Hamburg), Bernard Dionysus Geoghean (Berlin), Alfred Gottwaldt (Berlin), Karl Grosse (Berlin), David Gugerli (Zurich), Jan Hendrik Haack (Hamburg), Heiner Hautau (Hamburg), Anke te Heesen (Berlin), Hans Joachim Heins (Munich), Matthew Heins (Ann Arbor), Boris Hochfeld (Hamburg), Tom Holert (Cologne and Berlin), Manfred Holler (Munich and Hamburg), Gisela Hürlimann (Zurich), Sam Ignarski (London), Tim Jung (Hamburg), Christa Kamleithner (Vienna and Berlin), Susanne Kill (Berlin), Rainer Klose (Karlsruhe), Markus Krajewski (Weimar), Joachim Krausse (Berlin and Dessau), Jesse Lecavalier (New York), Christiane Leiska (Hamburg), Heidi Linsmayer (San Francisco), Marc Matter (Düsseldorf), Philipp Messner (Zurich), Gesa Mueller von der Haegen (Karlsruhe), Gerhardt Muller (New York), Ariane Pauls (Berlin), Robert Pfeffer (Cologne), Nils Plath (Berlin), Walter Prigge (Dessau), Karla Reimert (Berlin), Tilman Rhode-Jüchtern (Jena), Sarah Sander (Linz), Gabriele Schabacher (Cologne), Armin Schäfer (Munich/Hagen), Martin Schmittseifer (Cologne), Samy Schneider (Hamburg), Manuel Schubbe (Berlin), Jeanette Schuster (Stuttgart), Lothar Schuster (Berlin), Bettina Sebek (London), Allan Sekula, Olaf Sobczak (Hamburg), Torsten Stracke (Hamburg), Theo Thiesmeier (Berlin), Ingo Timm (Bremen), Daniel Tyradellis (Berlin), Gerhard Vinken (Aachen), Cornelia Vismann, Jutta Wangemann (Berlin), Timon Wehnert (Berlin), Corinna Westrich (Hamburg), Eva Wolff (Cologne), and Christoph Zschaber (Berlin), as well as those who took part in my seminars on the subject at the Universities of Weimar, Dessau, and Karlsruhe and the participants in the World from the

Container workshop in the Prater of the Volksbühne Berlin in June 2005.

The publication of this book for English readers was initiated by Paul N. Edwards and Geoffrey C. Bowker, whom I want to thank. It is an honor for me to be part of their new infrastructures series at the MIT Press. The translation was financed through funding by Börsenverein des Deutschen Buchhandels.

Accident of MSC *Napoli* container ship in January 2007. Credit: PA Images. Used with permission.

Introduction: The Accident

On the morning of January 18, 2007, nearly 50 miles south of the headlands of Cornwall, a container ship sailing under the British flag in the English Channel was in distress. MSC *Napoli* was at the very beginning of its journey from Belgium to South Africa via the Portuguese city of Sines when it crossed the destructive path of Hurricane Kyrill. Amid 40-foot-high waves and a 70-mile-per-hour wind, the 26 members of the crew were successfully rescued by helicopter. The disabled freighter was initially supposed to be towed to the nearest harbor, but instead it was put aground off the coast of Cornwall because it was threatening to break apart.

As a result of the continuing poor weather and an alarming starboard tilt of 35 degrees, 116 containers spilled into the sea. The remaining 846 containers stacked on the deck were able to be unloaded by March 9 with the help of a floating pontoon. The recovery of the approximately 1,300 containers stacked below the deck dragged out for months. Of the steel boxes that had gone overboard, 73 washed up along the coast and 11 were spotted on the sea floor. The rest were considered missing.

With a capacity of 4,419 TEUs (twenty-foot equivalent units—the internationally determined designation for 20-foot standard transport containers), MSC *Napoli* was a midsize container ship. When it was launched in 1991, under the name CMA CGM *Normandie*, it was the first fully rigged post-Panamax container ship—that is, the first container ship that was too large (902 feet long and 122 feet wide) to cross the Panama Canal.

A comparison with the dimensions of today's largest container ships reveals the monstrous growth that remains nearly unchecked in the transport sector, compared to any other industry, despite oil crises, collapses in the capital market, warming climate, and shipwrecks. The French shipping company CMA CGM's *Marco Polo* loads up to 16,000 TEUs and measures nearly 1,300 by 180 feet. Produced on order from the world's largest shipping carrier, the Danish firm A. P. Møller Maersk, the new Triple E Class ships, 1,312 feet long and 194 feet wide and with a loading capacity of up to 18,000 TEUs, set sail in 2013. What will happen if a ship of such proportions loses even a fraction of its payload?

The wreck of MSC *Napoli* off the coast of Cornwall in 2007 caused a state of emergency. In earlier times the area was notorious as a home for wreckers, who lured their victims into the shallow coastal waters with false beacons to then plunder the ships run aground. When a few dozen containers washed ashore on the beach of the small, quiet, coastal town of Branscombe, inhabited primarily by wealthy retirees, thousands of people traveled from across England to grab a bit of loot. They besieged the beach. The narrow village street was blocked, with parked cars stretching for miles from the site. The collectors and looters weren't stopped by warnings that the containers could be carrying dangerous cargo.

Media from around the world reported extensively on the "Night of the Treasure Hunters." "Hundreds of Beachcombers

Beachcombers in Branscombe. © picture-alliance/dpa—Report.

Pounce on Cargo," "Scavengers in the Hunt for Booty," and "Dreams Run Aground" read the headlines of German daily papers. The "self-service party" and "late Christmas" were topics of discussion. The beach was said to be transformed into a "supermarket," littered with gears, steering wheels, and other spare car parts, along with wine barrels, cookie tins, first-aid kits, perfume bottles, sneakers, diapers from Arabia, shoes from Cyprus, empty French barrels meant for South African wine, dog food, clothing, household appliances, and toys. Even a tractor washed up.

Members of an immigrant family from New Zealand said that they recognized the personal effects of their European grandparents among the suitcases and furniture that were dragged off as the cameras rolled. In *Die Zeit* newspaper on February 1, 2007, one of the family members reported the following:

My parents were sitting at breakfast when a journalist called and said that our containers had been found. We turned on the TV and watched as someone with an iron rod beat a hole in the container. Then looters took out our things, memorabilia from my late grandparents: wedding photos, an old table where we always sat with them, a sofa.

Also causing a furor were the images of young men taking off on brand-new BMW motorcycles retrieved from containers. But more than the feelings of outrage, anxiety, and envy, a fascination dominated: For once, the boxes that were circulating in all parts of the world in such incalculable numbers had exposed their contents. For once, those hermetic boxes, which usually concealed their cargo even from the ship's crew and the dockworkers, had opened!

According to the insurer's figures, only five containers had broken open and been looted on the beach at Branscombe. But the number belies their importance. What the containers offered the travelers on the scene as well as the public around the world was a cross-section of the present state of world culture, of global consumer capitalism.

In the last 50 years, containers have done more than bring about a fundamental change in the transport of goods on sea and on land; they have reformed beyond recognition the culture of the harbor, which had endured for thousands of years. Containers have supported significantly the emergence of a system of production and consumption that circles the globe, leaving almost no place on Earth untouched. This system has been discussed for several years with equal parts intensity and controversy under the name *globalization*. The boxes are the core and the crowning element of a logic of modularization and optimized distribution called *logistics*, which, since the nineteenth century and with considerable acceleration in the twentieth century, has successfully moved from the factories and the battlefields into all sectors of society.

Modernity, as a specific and systematic form of organization of social life, is subject to a logistical structure, an operational order of knowledge, and to date the container has proved to be the most successful material agent of this logistical access to the world. (The most successful "immaterial" agents are computer programs). The system of the container structures and encodes everything that falls within the realm of its process—that is, nearly everything. It establishes its own spatial and temporal order. It transforms the world into a moving warehouse and arranges it in the mode of standardized movable spatial units, switched processes, and clocked times.

If "the furrow drawn in the ground by the plow, and the (grain) silo" may be seen as "archaic technologies of hominization," as media scientist Bernhard Siegert has written, then the container is the modern answer to the ancient question of cultivation and utilization that constitutes culture.[1] Containerization is a prevailing cultural technology of the 20th and early 21st centuries. Thus, an investigation of the principle of containerization cannot restrict itself to the more narrow realm of the transportation of goods and logistics from which the system of standard containers emerged.

Rather, one must also address the effects of the concept in other areas of knowledge and cultural practices. It seems that containers play as decisive a role in the organization of people, programs, and information as they do in that of goods. They not only physically appear in every imaginable place in the city (such as subway stops and airports) and in rural areas, they also appear in such cultural domains as architecture and urban planning, psychology, philosophy, pedagogy, business administration, communications and information, film, television, theater, and art.

These environments, which the container intruded in and which have been reconfigured or newly created altogether, are the subject of this book. Each chapter begins with a specific "container situation" and makes excursions from there into the world of transportation history, logistics, architecture, information technology, thought, and the organization of material, people, and knowledge.

Looted containers on Branscombe Beach, January 2007. © picture-alliance/dpa—Report.

1 Time Capsules

Clearly, the container is part of the message.
—Robert Ascher

In the story of Branscombe's stranded containers (see Introduction), the steel box with the physical legacy of a New Zealand emigrant family's late grandparents opens up an additional temporal dimension. Rather than being new goods from around the world, whose individual stories as objects of use still lie before them, these goods indicate a past that has already been lived. The material evidence of multiple individual biographies was sent on the journey in order to continue their stories somewhere else in the world.

Instead, the goods are stranded, and now they're reaching the wrong address. Because their means of transport had an accident, their protective milieu has been destroyed and their individual historical context has been dissolved. Pictures of this process are being carried worldwide on television. In reality, as a faceless and ahistorical standardized link in the endless chain of intercontinental goods transport, the desecrated emigrant container is reminiscent of overseas boxes exhibited in shipping and emigration museums. A twofold distance has come about between the current presence of "recovery," or the public plundering of memorabilia, and their previous living presence: that of the deaths of the objects' former owners and that of the misguided arrival.

Andy Warhol, Westinghouse, and the Crypt of Civilization

"What you should do is get a box for a month, and drop everything in it and at the end of the month lock it up. Then date it and send it over to Jersey."[1] Since 1974, Andy Warhol, who constantly struggled with a need for space as a notorious collector of (pop) cultural artifacts, handled a portion of the daily flood of incoming objects as follows: He sifted through them, then he tossed whatever was worth keeping, but not acutely important, into a standard brown box, which sat on his desk for this purpose. Once a month, the boxes were sealed, dated with a colored sticker, and stored. He ironically called the box objects Time Capsules, and he initially conceived of their creation less as art than as part of a refined logistics of order with which he fooled himself.

By the time of his death, 612 of these boxes had been created, containing correspondence, photos, newspaper clippings, books, magazines and journals, receipts, invitations and flyers, disgusting things like leftover pizza, extremely valuable things like Warhol's early artworks, and oddities like Clark Gable's shoes, Concorde utensils from Air France, and a seventeenth-century German-language book about wrestling.

If the existence and the tremendous historical potential of the Time Capsules were mostly unknown during Warhol's lifetime, today they are understood as an integral component of the artist's attempts to capture the fleeting nature of passing

Time Capsules

Andy Warhol, Time Capsule. © 2014 The Andy Warhol Foundation for the Visual Arts, Inc. / Artists Rights Society (ARS), New York.

time and human experience in serial works. An entire section of the Andy Warhol Museum in Pittsburgh is concerned with opening these boxes that have traveled through time, cataloging and conserving their contents piece by piece. Conservators at work can be seen through a glass wall, and they will continue for quite a while—until 2009, only about 200 of the more than 600 boxes have been opened. Many of their sometimes valuable and highly sought-after contents will remain untouched for years, sent into latency according to Warhol's "store away" concept, by which he got rid of things and preserved them at the same time. A museum colleague compares the archivists' work to "a kind of relentless opening of Tutankhamen's grave."[2] Warhol's Time Capsules are (post)modern temporary burial chambers.

Time capsules are conceived of as an "archaeology for the future," as G. Edward Pendray, the director of the Westinghouse Time Capsule, one of the largest time capsule projects between the world wars, noted in a 1938 article.[3] They gather and enclose an inventory of artifacts and documents of a culture, from which future researchers are supposed to be able to reconstruct the nature and functions of an age long since passed.

Although the first modern time capsule, with a life span of 100 years, was started in 1876 at the World's Fair in Philadelphia, the golden age of the great time capsule projects began only in the late 1930s and lasted until the early 1980s, according to archivist and time capsule historian William E. Jarvis. Thus, it corresponds rather precisely with the period from the beginning of World War II to the end of the strongest phase of the Cold War. On the one hand, time capsules are an attempt to create a link to tradition that will not be broken by a global catastrophe. On the other hand, they serve as a reflection of their own culture from an assumed geological or extraterrestrial (that is, temporally and/or spatially distant) standpoint.

In 1938, the preparations for one of the most ambitious time capsule projects of all time, the Westinghouse Time Capsule of Cupaloy, were in full swing. The torpedo-shaped time capsule, about eight feet long and eight inches in diameter, was supposed to send a compressed cross-section of (American) civilization into a future 4,000 years removed. The time-travel container was constructed like a thermos. Its hull consisted of a copper,

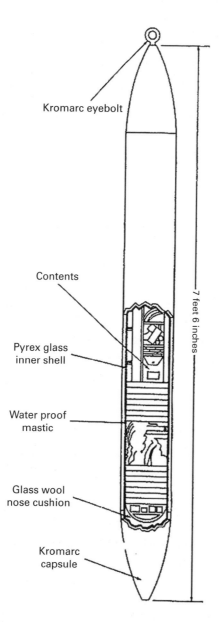

Schematic design of the Westinghouse Time Capsule. Courtesy of the Detre Library and Archives, Senator John Heinz History Center, Westinghouse collection.

chromium, and silver alloy. Inside was a sealed capsule of hardened glass, into which things from five different categories were placed: small objects of daily use, such as alarm clocks, nail files, playing cards, and toothbrushes; a microfilm essay with photos and lexical entries; a newsreel with characteristic scenes of the 1930s in sound film; materials of the time, consisting of swatches and samples from renowned American companies; and miscellaneous things like money, seeds, and print type.

Time capsule projects are meant to send a compressed vision of their time, a kind of snapshot of global development. In the teeming variety of objects and events, patterns can be recognized only from a certain distance. What is relevant and what is not is often judged entirely differently at a later point in time. This is precisely why time capsules do not limit themselves to current conceptions of high or mainstream culture but also include the fleeting, the arbitrary, and the accidental, evidence of everyday and popular culture. This insight into the importance of preserving the fleeting—quite comparable to Warhol's Time Capsules—is brought to bear in all time capsules.

This is particularly true for the other grand time capsule of the 1930s, Crypt of Civilization, a project of Oglethorpe University in Atlanta, Georgia. It is not a capsule in the narrow sense of the word; rather, it is a crypt lined and tiled with steel walls 20 feet long, 10 feet wide, and 10 feet high (corresponding, more or less, to the measurements of a 20-foot container), with a volume more than 1,000 times that of the Westinghouse capsule. The leader of this time capsule project, Dr. Thornwell Jacobs, the president of Oglethorpe University, assembled a collection of the achievements of art, science, and industry in the first third of the twentieth century that should serve as an academic resource for scientific research in the eighty-second century.

Crypt of Civilization includes a 640,000-page comprehensive collection of newspapers, magazines, books, patents, photographs, and other items on microfilm; films of historical events since 1898; and a film projector as well as a windmill to generate electrical current for it. There are also many hundreds of objects of daily use, such as a model train set, a pair of male and female mannequins in a glass display case, a pair of glasses, a radio receiver, a glass refrigerator container with lid, a plastic flute, a pair of women's hosiery, an illuminated vanity mirror, phonograph records, and two early television receivers.

Shipments from the Past

These modern time capsule projects are planned well in advance. They all operate with materials and forms that are designed for the long durability of the capsules. However, a container is not necessarily consciously assembled, loaded, closed, sealed, and sent on a journey through time in order to function as a time capsule upon its arrival. It also need not be a container in the proper sense of the word. There are accidental encapsulations, stemming from misfortune: objects unintentionally vanished into the depths of space and time that reveal themselves as time capsules only upon their recovery—for instance, extinct cities, abandoned homes, scrapped automobiles, or sunken ships.

The archaeologist Robert Ascher, engaging with the rampant time capsule mania of the day in a 1972 lecture at Ithaca College, New York, entitled "How to Build a Time Capsule," emphasized that only these "unplanned messages" could be real objects of archaeological research, because "planned messages" give the illusion of being "an archaeological excavation in the future."[4]

Upon closer inspection, however, Ascher noted that it becomes clear that the objects of real archaeological examination, as remnants and traces of actual lives, reveal a *slice* in time because each object has its own history, which is the history of its users and its uses. In contrast, most modern planned capsules conserve only a *point* in time, since they lack the level of lived history with their new, unused objects.

Pompeii, the ancient Italian city obliterated in a moment on August 24, 79 CE, by the eruption of the volcano Vesuvius and frozen beneath a layer of ash, lava, and pumice, can be seen as a paradigm for the unplanned time capsules arising from accidents and coincidence. A catastrophe abruptly ended the city's existence in its own historical time, and a circumstance that can only be termed "fortunate" in hindsight ensured that the city's final condition at the time of the catastrophe was conserved. From the manner and arrangement of materials carried into the light of another, later time, posterity can now draw conclusions, such as when one tries to reconstruct the structure of a game interrupted suddenly—its goal, its rules, its process, based on the construction of the playing field—from the number and arrangement of the figures.

The excavated city and the salvaged ship are transmissions from another era. Their arrival sometimes brings together periods far removed from each other. For archaeology, which has taken shape since the eighteenth century, the gradually excavated city of Pompeii presents itself as an inventory of ancient urban culture: a shipment from the distant past that has served, since the Renaissance, as the beginning of European history. However, it was a shipment whose address was uncertain until its arrival: it was only on April 6, 1748, as the Spanish engineering officer Colonel Roque Joaquín de Alcubierre began the excavations with the permission of the Neapolitan royal family, that the buried city found its destination as an ancient legacy. Only after that date did Pompeii become a time capsule—nearly 1,700 years after its downfall.

What kind of time capsule would a similar catastrophe produce today? If global climate change takes its worst projected path, the polar ice caps will melt away, the waters will rise, most of the current coastal regions will be lost, and, within 200 years, vast portions of the continents will become uninhabitable as a result of the increasingly extreme climate conditions. As emigration into space will most likely still not be a viable option in the twenty-third century, the colonization of the world's oceans through the development of the seabed might begin for an aquatic or a subaquatic existence. In the course of this process, more and more of the thousands and thousands of shipping containers that went overboard from ships in the twentieth and twenty-first centuries will be discovered. In many parts of the ocean that have seen particularly fierce storms, there are entire fields of containers.

"Unprecedented opportunities for fields of research in future underwater archaeology have been thrown overboard," ethnologist Konrad Köstlin writes, "and these objects—as nineteenth century historian and king of source materials Johann Gustav Droysen called them—are 'frozen in thought, so to speak.' The containers are just lying there, tossed overboard in stormy seas, and waiting ... for future research."[5]

What will the rediscovered steel containers, these mobile warehouses of modern consumer culture, represent to future researchers? Presumably, the situation will be similar to that of the archivists now in the Andy Warhol Museum. From the boxes, the researchers will be inundated with a seemingly endless supply of the most varied objects, evidence of "high" and "low" culture, confusing in its arbitrariness. But with an overall perspective of the contents of a multitude of containers, patterns could be recognized and conclusions drawn, and the inventory of the contents of a few hundred or thousand containers would yield a relatively complete inventory of the material culture of our time. The containers would function as time capsules, like the emigrant container at Branscombe or the ruins of Pompeii.

The World from the Container

In September 2001, New York journalist and author Richard Pollak undertook a journey by container ship from Hong Kong to New York. He begins his description of the trip on board the *Colombo Bay* with an inventory of his luggage, which he spreads out piece by piece in the ship's cabin to examine the origin of each object and each piece of clothing:

If I needed any evidence that containers play a central role in my own life, it now spills into the cabin. One blue Brooks Brothers button-down shirt, made in Thailand; gray dress slacks, from India; blue Helly Hansen rain slicker, Sri Lanka; blue rubber rain pants, Taiwan; red baseball cap, China; Panasonic CD player, Japan; CD pouch, China, Korea, Philippines, or Indonesia, take your choice; Bell and Sony tape recorders, China; Panasonic tapes, Japan; Olympus Infinity 5 camera, assembled in Hong Kong from parts made in Japan; Casio quartz travel clock, assembled in Thailand; Sanford Uni-Ball Onyx micro pens, Japan; IBM ThinkPad AC adaptor, China; laptop carrying case, Indonesia; shoulder bag, Korea; garment bag, Taiwan. The red bathing suit, knit gloves, watch cap, Samsonite toilet kit, socks, yellow highlighter pens, and swimming goggles bear no indication of their provenance, but odds are that at least half these items were made in Asia, too. "What most North Americans don't understand," Jeremy [Nixon, a high-level executive at a major shipping line] has said in one of our first talks about this voyage, "is how reliant they have become on Asian products." And on imports from elsewhere: ThinkPad and its power cord and mouse, and Gillette razors, Mexico; khaki shorts, El Salvador; gray shorts, Guatemala; Van Heusen button-down shirt and Jockey shorts, Costa Rica; blue dress jacket and blue denim shirt, Canada; Sanita clogs, Denmark; blue woolen Lands' End sweater, Scotland. I made no effort to categorize items I didn't bring with me, but for the record the aforementioned Lux soap was made in Indonesia and the *Colombo Bay* herself, as noted, was made in Japan. Full disclosure: my Kodak film, Duracell batteries, Lands' End warm-up jacket, New Balance running shoes, black sweatpants, two belts, large brown suitcase, and Penguin paperback of *Moby-Dick* were Made in the USA.[6]

The realization forces itself upon us that we live in a world from the container. Tens of millions of standardized loading units are in motion every day on the tracks, streets, and seas, transporting the goods and intermediate goods of global consumer capitalism from one part of the world to another. In a certain way, every one of these packed, sealed, and sent containers is a time capsule: The transport goods disappear for a determined amount of time, beyond all control. The container is the sole object and subject of all processes concerning transport—the loading and unloading of ships, vans, or trains and the addressing, documentation, and customs procedures; it is the temporary, mobile medium of enclosure, the metacontainer.

Enclosed and sealed in the container, the cargo is subject only to the time-space conditions of the box. Thus, to a certain extent, the cargo exits the time-space continuum of the worlds in which it is produced, through which it is transported, and in which it is consumed. For this reason, every opening and unloading of a container is similar to a reemergence and resurrection. Container transport is the movement of the immobilized.

Consequently, the transport of perishable food, particularly meat and fish, is a good example. By freezing, all organic processes must be interrupted as quickly and completely as possible. For this, one needs cooling techniques and special transport containers. The deep-frozen foods are ripped from their own time (of decay) and made into an anachronism: timeless and spaceless in the duration of transport and storage. Only on the other end, in an entirely different place and with a temporal gap in their "bio-graphy" that can never be closed, do they reenter their own time.

Interrupted Shipments, Unplanned Time Capsules

The technical and organizational development of modern means of transport has robbed the arrival of shipments of their special qualities and turned them into a routine occurrence. Yet, paradoxically, the internationally standardized transport container, which is a high-end product of this technical and organizational development, can bring about precisely the opposite effect and (once again) fill the standard format with meaningful content, by making it lose its systematic purpose and become a time capsule.

The belief in tradition was largely lost in the twentieth century. Too many shipments are permanently under way—too much data, too many goods, and too many possible meanings for a shipment to simply have one destination/destiny and, from this, one story.

Under these conditions, the (entirely expected and predictable, altogether unfateful) arrival of a shipment can be regarded as a nonevent, whose only relation to an event lies in the fact that it *did* arrive, that it did not go astray. Conversely, the interrupted shipment, and thus the unplanned time capsule, becomes an event and consequently a possibility to produce continuity, or at least the effect of continuity, to yield a context of transmission that is constitutively interrupted. This is based on a shipment that *must* be diverted, obstructed, and stopped in its path in order to appear at an unexpected time at an unexpected place and to unfold a surplus of meaning, precisely because of the improbability of its arrival, that makes it stand out amid the myriad of other shipments.

The paradoxical (accidental) logistics of destination/destiny: shipments reach their goal according to plan but in fact never arrive because they were never sent. Instead, they remain in a systemic context of distributed production. The world is a gigantic assembly line, on which standardized production units are processed without interruption and for which there exists no principal difference between production and transport. Only the shipments that thwart the plan and do *not* reach their goal have a chance to arrive somewhere, because a destiny sent them by interrupting their systemic path. To the extent that being transported became the normal form of existence, (planned) shipments lost their special value.

Thus, the system of shipment, rather than the contents themselves, becomes a transcendental reference; the postal service, the transport logistics, and the communications network become a medium and an existential reassurance. It is not the shipment addressed to an individual that is decisive, but rather the possibility of being addressed; not the contents of the package, but its status as a shipment.

In the 2000 film *Cast Away*, an employee of FedEx (which supported the production of the film with $54 million) crashes over the ocean in his cargo airplane and is stranded on a desert island. Gradually, more and more FedEx packages wash up on the island from the hold of the doomed aircraft. The stranded

FedEx employee finds a use for nearly all the objects in the packages.

However, what turns out to be the most important symbol for the modern Robinson Crusoe (whose telling name is Chuck Noland) is the one package that he does *not* open: The standard box of the company—with the printed logo, sender and recipient addresses, postmarks and stamps—refers to the existence and the functioning of the logistics network itself and consequently allows the stranded man to believe in his own possible rescue. The dominance of the self-referentiality of this shipment is strengthened by the imprint of a winged figure (Hermes, or perhaps an angel) applied by the sender. The divine message here is not any kind of mailed contents, but rather the fact that there is a postal system or goods transport system at all.

Noland is ultimately found by a container ship on the high seas after he dares to flee the island with a homemade raft (whose sail consists of half of a standardized, industrially produced plastic toilet—another accidental shipment—that washed up one day and heralded the basic accessibility of the stranded recipient's home civilization). At the end of the film, Noland delivers the packet with the Hermes symbol to a farm somewhere in the middle of nowhere. Since he doesn't see anyone, he adds a handwritten message to it, saying, "This package saved my life. Thank you, Chuck Noland."

Transport companies make a profit by minimizing the time in transit and maximizing the turnover rate. Their vehicles are found either in spaces of transit (on transport routes on water, on land, or in the air) or in holds (terminals, distribution centers, or transfer stations)—places in which the world, or the randomness of events and the complexity of circumstances, appears as little as possible. In this way, an economy of containerization has come about in which places are sought out according to prices and production is distributed across very distant locations along the global supply chain, since distance plays only a very subordinate role. Most of the containers are filled not with end products but rather with the intermediate products of industrial production: screws, machines, toys, auto or electrical appliance parts, synthetic resins, or recycled paper.

The history of the end products produced on these global assembly lines, if one may even speak of such a thing, is always realized in multiple, successive sequences, according to a coincidental combinatory calculus. Which hair is attached to which Barbie doll head, for instance, or which radiator grill is added to which automobile is decided only very late and randomly in a particular phase of the production process. The American economic journalist Marc Levinson has researched the stations that a Barbie doll passes through before coming to the market in the United States: the bodies are fabricated in China through molds that come from the United States and with machines from Japan and Europe, the nylon hair is produced in Japan, the plastic her body is made of comes from Taiwan, the pigments that give her color come from the United States, and her clothing is made in China.

The time capsule projects position themselves against this sequential disintegration, or, to put it more positively, against this sequential constructivism of industrial production. But what kind of time is it in (or with) which one tries to send the world, and thus history, in containers? In the years immediately before World War II, when Pendray coined the term *time capsule* and the preparations for the first two large-scale time capsule projects were under way, standardized containers were already in heavy use on the railways and roads in the important industrial nations, such as the United States, England, Germany, France, and Belgium. It was also at this time, the story goes, that the later inventor of the modern freight container, Malcom McLean, conceived the idea of intermodal container transport between trucks and ships.

It is perhaps no coincidence that such supposedly disparate ideas came about simultaneously—namely, the use of standardized containers for more efficient transport across large distances and with varied modes of transport, in order to make goods cheaper and the profits gained from them higher, and the implementation of highly specialized containers for transport through time, in order to conserve a level of knowledge, culture, and technology for future humans. More to the point, one might say that precisely those transport containers that provide for a continually faster decay of the present, on the one hand, open up, on the other hand, spaces for preservation and passing on. As containers, they transport the daily fashions; as time capsules, they promote thinking in large periods, which transcends the short-term system of modernity and which should work as a therapeutic agent against the "dromological revolution" of modern culture, as Paul Virilio, a great theoretician of modern speed and misfortune puts it—against the permanent catastrophe of disappearance.[7]

Accident Archaeology

In storms and shipwrecks, containers regularly go overboard. Just how many, no one knows exactly; shipping lines as well as insurance companies keep these figures under wraps. Estimates are several thousand every year. Measured against the total number of containers, this is an infinitesimally small figure: every day, there are about 10 million of them on the move simultaneously, and the 2007 year-end total was 400 to 450 million. Yet the lost boxes are slowly growing to be a problem, not only because of the possible pollution of the oceans by dangerous substances or the costs arising from the loss of cargo, but also because the boxes often do not sink immediately, instead bobbing like an iceberg or a giant message in a bottle just beneath the water's surface.

This can have fatal repercussions for smaller ships in the event of a collision. In June 2003, the Flügge family from Bremen felt its 56-foot-long yacht *Monsun* collide with an object floating in the sea, several hundred miles off the coast of Newfoundland. The boat slowed down with a jerk. A mast broke, and water entered the hull in several places. A half hour later, the crew had to abandon the wooden ship, which had just recently been restored, and it sank in the cold waters of the North Atlantic.

Although it cannot be said definitively what brought the *Monsun* to the seafloor—it could have been a sleeping whale or a floating tree trunk, for instance—the force of the impact and the severity of the damage indicate a container. Nevertheless, opinions vary widely about how great the danger is of colliding with a container at sea. Whereas some experts are convinced that the chances are incomparably higher of colliding with a tree trunk falling from a truck on the highway, others emphasize the threat stemming from the ever increasing number of containers transported and thus also lost, especially as the intensity of storms in the North Atlantic and North Pacific has increased in the last 30 years.

What does this new potential for collision at sea tell us? Two very basic themes are expressed in the container accident. The first is that this shows a mixture of (industrial) culture and nature typical for the modern world, to which people in densely populated nations have become more and more acclimated since the nineteenth century. But it has been perceived and described on a broad basis only in recent times—namely, since the advent of global observation technologies—in the so-called last wildernesses, the vast deserts and mountains, the oceans and the polar regions. If there is an equal likelihood of colliding with a container in the ocean as there is of colliding with a sleeping whale, then something is happening on a magnitude thoroughly comparable to global climate change.

The second theme is the technology affected by the accident itself, which appears in a new light in its informative dimension precisely because of the accident. I would like to begin with this aspect.

The world of containers suggests a smooth, lossless processing of transportation, almost immaterial in its textbook implementation of a logistical ideal—so efficient, measured by transport volumes, that the transportation costs could become a negligible factor. From this point of view it is easily forgotten that this is nevertheless a matter of heavy metal, a gigantic technological system of steel and silicon that requires tens of thousands of human workers to function.

However, in the accident, the facade shatters, the panels fall, and the housing bursts. The machinery is temporarily opened up, and its wheelworks, circuits, and networks become visible; the political and financial contexts, the legal foundations, the social organization, and the ethical decisions are laid bare. How much is a human life worth? How much is a nonhuman life worth? How great a price is society prepared to pay for the functioning of a technology? How are progress and security balanced?

Contrary to common perception, the accident does not represent the exception for a technology but rather is as much a part of it as its functioning is. Every technology is bound to a specific form of accident. Paul Virilio formulates the situation this way: "To invent the sailing ship or steamer is to invent the shipwreck. To invent the train is to invent the rail accident of derailment."[8] Technology and accidents create each other.

Since the beginning of the twentieth century, daily life has become a "kaleidoscope of incidents and accidents," as Virilio writes. In keeping with this, he advocates "a new kind of museology and museography: one which consists in exposing or exhibiting the accident."[9] The new type of accident that the container ship brought in its wake lies in the containers gone overboard, the deployment of time capsules or "time bombs" (here the original term, abandoned by G. Edward Pendray, becomes surprisingly topical).

The misfortune of MSC *Napoli* merits inclusion in Virilio's museum of accidents in two ways. First, each lost container has the potential to bring about a variety of incidents and accidents: pollution of the environment and poisoning of people through transported hazardous materials; containers causing collisions with ships; loss of the transported goods, whether a container cannot be found or recovered, it breaks open and loses its cargo, or it is plundered.

In addition, the subject of shipwreck, its ramifications and its potential causes, provides an impression of the extremely diversified practices of international container transport. One need only think of the shipping company, the owner, the charterer, the possibility of flagging out a ship, the complexities of insuring a ship, containers and container cargo, and so on. In sum, one gains the impression of a vast distribution system in which not only cargoes but also jurisdictions and responsibilities change constantly.

Thus, the owner of MSC *Napoli*, built in 1991 in a Korean shipyard, is an English firm based in the Virgin Islands. It is commercially and technically supervised ("operated") by a London ocean carrier, its home port is London, and it sails under the British flag, but it is chartered by a Swiss shipping company, the Mediterranean Shipping Company (MSC). This company's fleet consists largely of ships that have already sailed in service to other lines, and their names and paint often conceal colorful histories. (MSC also operates its own new ships, which can be differentiated by their names; the new ones bear women's names, like MSC *Stella* and MSC *Loretta*, whereas the charters are named after cities, nations, and other parts of the world.)

MSC *Napoli* had already had multiple names and clients. The vessel left the shipyard in 1991 as CGM *Normandie*. In 2000, it became known as *Nedlloyd Normandie*. In 2001, once again under the control of its previous charterer, Compagnie Générale Maritime (CGM), which had merged in the interim with another French company, it became CMA CGM *Normandie*. In 2004 it became part of MSC's fleet, bearing the name that would make it world famous in January 2007.

The ship had already had a series of accidents. In April 2001, shortly after it returned to French hands, it ran aground on a coral reef during a passing maneuver in the Strait of Malacca and lay stranded there for 60 days before it was towed to a harbor in Vietnam and repaired. In December 2001, only two months after it was brought back into service, it suffered still more damage to its hull when it crashed into a quay wall in the port of the Saudi Arabian city of Jeddah.

In August 2002, it ran aground once more, again in Jeddah, but only minor damage was caused in this case. Speculation that cracks on the deck, which could have contributed to the breaking of the hull, might have originated from earlier accidents were not confirmed, however. Instead, the final report of the British Marine Accident Investigation Branch revealed that a construction error that could affect other ships of the same kind led to the fatal rupture in the hull that caused the ship to capsize in the English Channel.

On the Trail of Lost Containers

What becomes of the containers that disappear? If they reappear, as on the beach at Branscombe, then the matter is relatively simple. Each standard ISO container carries, on each of its four sides, a code clearly identifying it. The codes for all containers worldwide are distributed by the Bureau International des Containers (BIC). In the cargo shippers' databases, the histories and the current cargo of the containers are recorded, accessible by their respective codes. Thus, it can be determined what was in each container and for whom it was intended. As long as the contents of a container have not been plundered outright, they can be delivered to the recipient, with some delay, even after an accident.

Naturally, it is more difficult when a container breaks open on the high seas, leaving its freight to the whims of the wind and the waves. But it is precisely these cases that have become the specialty of a peculiar scientific method in the last 250 years: ocean current research. This leads me to the first dimension of the container accident mentioned previously, the mixture of culture and nature. In May 1990, the *Hansa Carrier* lost 21 containers in a Pacific storm north of Japan, and four of the containers broke open. The payload, more than 60,000 pairs of Nike sneakers, poured out into the sea. About nine months after the accident, 1,600 sneakers washed up along the coast of Oregon, on the Queen Charlotte Islands in Canada, and on other beaches along the northern Pacific coast.

Curtis C. Ebbesmeyer, an oceanographer based in Seattle who studies the ocean's currents with the help of a global network of jetsam hunters, documented the path of these sneakers. He worked with his colleague James Ingraham, who wrote a program to simulate surface flow (Ocean Surface Current Simulations, or OSCURS). With this program, they calculated the probable routes of the drifting shoes and compared their results with the actual routes of the shoes that had washed ashore, been reported, and identified by their serial numbers.

Twenty months after the loss of the *Hansa Carrier*, another ship found itself in a severe storm in the Pacific, this time southeast of the Aleutian Islands. The shipping company revealed the precise geographical location of the accident to the oceanographers only on the condition that the name of the ship would be kept to themselves. This issue points out one of the prime difficulties with this type of research: it relies on data—namely, the precise coordinates of accident sites—that the shipping operators would prefer to keep secret.

Twelve 40-foot containers went overboard. One contained 29,000 units of bathroom toys: blue turtles, yellow rubber duckies, red beavers, and green frogs, all made in China. In November 1992, 10 months after they washed overboard, the first toys turned up on the coast of Alaska. After two years, several plastic animals were sighted in the ice of the Bering Sea before they continued their journey with the thaw. Ten years later, toy animals

Time Capsules

Accidental current probe.

were still reaching land, some far to the south all the way to Hawaii.

And these aren't the only "inaccurate drifters" (in the jargon of ocean current researchers) that have been set loose by accidents in the last 20 years. Indeed, the fleet of cargo converted to "probes" also includes 100,000 toy cars, 34,000 hockey gloves, several hundred thousand hermetically sealed Riesen chocolates, half a million beer cans, 3.9 million Lego figures, and, once again, 39,000 Nike sneakers. Though released unintentionally, the goods are fed into the system of the oceans like data in the oceanographers' simulation program. By comparing the results of these two experimental systems, which record the migration of lost merchandise through the correlation of data from three transport routes—cargo shipping paths, surface currents, and wind direction—an increasingly precise map of Pacific

Ocean currents and an entirely new map of the global transport of goods are created.

A pioneer of this kind of scientific current research was King Albert I of Monaco (1848–1922). In 1885, he threw 1,675 buoys into the Atlantic along several lines between Europe and North America in order to investigate the path and speed of the Gulf Stream. Érik Orsenna tells this story in his 2010 sea tale *A Portrait of the Gulf Stream: In Praise of Currents*. Each buoy contained a message, drafted in seven languages within a sealed glass tube, requesting that those who found it should return the note "to the proper offices of their home country, so that it may be passed on to the French government, indicating the circumstances in which this document was found."[10] The buoys were found along the coast of Europe to Gibraltar, on the coast of Africa to the Canary Islands and Cape Verde, in the Antilles, and along the coast of Central America.

Since Albert I, an important branch of oceanography has developed from the practice of setting buoys. It has become evident in recent years that accidental current probes from damaged containers are often even superior to intentional buoys and scientific probes produced specially for this purpose and outfitted with all manner of recording and radio technology. On the one hand, the growing mass of cultural flotsam poses a threat to the ocean ecosystem that must be taken seriously—not to mention the environmental catastrophes regularly brought about by tanker accidents—but on the other hand, the inaccurate drifters contribute fundamentally to the knowledge of this system's function and thereby present possibilities for its salvation.

In the modern period, nature has become an integral part of giant mechanical networks, whose individual components and local effects are built and programmed by humans but whose

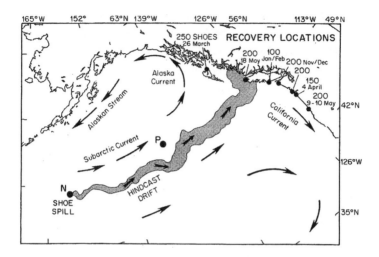

Shoe spill in the North Pacific. Map by Curtis C. Ebbesmeyer and James Ingraham Jr. From C. C. Ebbesmeyer and W. J. Ingraham Jr., "Shoe Spill in the North Pacific," *Eos, Transactions American Geophysical Union* 73, no. 34 (August 25, 1992): 361–365, doi:10.1029/91EO10273. Reproduced with permission of John Wiley & Sons, Inc.

global effects evade human control. With increasing frequency, nature strikes back with catastrophes. Hans-Jörg Rheinberger, a renowned philosopher of science and the former director of the Max Planck Institute for the History of Science in Berlin, writes the following:

> Today, at least at our latitude, nature ... has long since taken on the dimension of a technological network of things, a branching being, no longer unspoiled in any of its elements, a hybrid. Today, every ecosystem has technological aspects; perhaps it will be most important to recognize that, inversely, no technological system can be viably maintained in the long term without ecological components.[11]

Nature and culture form hybrid compounds, a network of human actions, representations, artifacts, and ecosystem conditions that mutually influence and illuminate each other. Insight into the function of the ocean ecosystem can be gained from plastic objects that have fallen into the middle of the ocean from containers. Where does the dimension of insight lie for containers that have fallen out of economic circulation as a result of the violent intervention of the forces of nature? What does the arrival of a container shipment mean? What can stranded containers say that cannot be inferred from their counterparts in circulation?

Accidental history is a history of technologically packed and timed sendings that become a shipment—that is, a destiny only when an accident happens. It is a history whose spirit is inspired by nothing but statistics, but whose accidents are nevertheless interruptions with a potential for insight that shows the presence of the everyday, the technological, and the commodity in global consumer capitalism.

Thus, the principles of global transit become visible: this transit cares little about local conditions, currents and storms, day and night, or winter and summer; rather, it draws its lines in accordance with its own system time, crisscrossing the globe. It continues a trend that began with the introduction of standard postage and that was taken to the extreme with the electronic shipment of information across global data lines—it levels space. It dissolves the topography and geography of the currents and winds, the rising and falling, the favorable and unfavorable environmental conditions, in support of a topology, an analysis of the connections between marketplaces, without consideration of real material circumstances. Ideally, the containers behave like bits, in the sense that they are processed over land

and sea as though on data lines. Their number is so large that the loss of a few boxes, even if each of them is its own world, a treasure chest or a museum, is of no further consequence. Nothing individual counts here; only the serial and the cumulative matter—boxes stacked in boxes.

In the moment of an accident, of no-longer-functioning, a system assumes, to a certain extent, an archaeological position with respect to itself: a time-space difference arises because something must be rescued, raised, and excavated. In this way, the accident becomes an epistemological constellation. This is the case not only for actual accidents—like doomed ships, wrecked trains and trucks, and sunken containers—but also for all historical moments in which connections between systems are not successful. An improvement does not take place, just as with all the constellations of misappropriation, of nonsystem compatible use, of the dissident implementation of transport containers. Every chapter of this book begins with such a found container, or a "container situation," which can then be interpreted and recontextualized as a shipment.

Accidents open space for questions concerning not only the technological but also the discursive preconditions of functioning or nonfunctioning or deviation from the model of operationality. What are the requirements for thinking about containerization? What order of things and of discourses accompanies an age of containers—of standardized reservoirs as an element of order, storage, and transport—or is produced by it? What lines of development in the history of technology cross in the system of containerization? How, and under what circumstances, did the fusion of transport and logistics come about? What are the subjects and objects of transport, and how are they constituted?

According to my initial hypothesis, in the twentieth century a basic logistical order emerged and enforced itself everywhere. The container developed, parallel to the computer, as a central element and the most logical implementation of this principle of a rationalized economy of transport or distribution. This book visits the container at the historical points where this principle was developed and where it differentiated itself, multiplied, and trickled into the most varied realms of society. It pursues it to those points at which new uses arise, where the container is reformed or broken open, perhaps announcing something like the end of the era of containerization.

DER BEHÄLTER

Offizielle Zeitschrift des Internationalen Behälter-Büros
bei der Internationalen Handelskammer

Nr. 1	Erscheint dreisprachig Deutsch Englisch Französisch	JANUAR 1934

Verlag : **Internationales Behälter-Büro,** beim Generalsekretariat der Internationalen Handelskammer
PARIS, 8e, 38, Cours Albert 1er
Fernsprecher : Elysées 62.42, 62.56   Télégrammanschrift : Containers-Paris-86

Nachdruck nur mit genauer Quellenangabe gestattet.

Was ist ein BEHÄLTER
von S. E. Silvio Crespi

Das Wort « Behälter » (cadre oder container) bezeichnet im weitesten Sinne alles, was Gegenstände irgendwelcher Art entsteht. So wurden z.B. Kisten schon seit dem fernsten Altertum als Beförderungsmittel benutzt. In solchen Kisten wurden auch die wilden Tiere für die römischen Arenen befördert.

In jüngster Zeit hat man für den Paket- und Gepäckverkehr zwischen England und dem Festland durch Kräne umzuladende Kisten benutzt, die mit der Eisenbahn zur Küste und zu Schiff über den Ärmelkanal befördert werden. Übrigens werden schon seit langem Möbelwagen oder Liftvans zum Transport von Umzugsgut über grosse Entfernungen benutzt.

Der moderne Behälter, wie er dem Weltmotorkongress in Rom im September 1928

vorgeschwebt hat, soll als Beförderungsmittel dienen und die Verbindung zwischen Eisenbahn und Kraftwagen herstellen ; er ist keine gewöhnliche Kiste, sondern der bewegliche, in Untereinheiten aufgeteilte Kasten des Waggons. Der Behälter in diesem Sinn teilt den Waggon und den Lastkraftwagen in zwei Teile : erstens einen Teil, den wir den festen Teil nennen werden und der von den Rädern, den Federn und der Plattform gebildet wird und zweitens den Kasten; dieser Kasten bildet den beweglichen Teil, lässt sich von dem festen Teil loslösen und kann ohne weiteres von einem Beförderungsmittel auf ein anderes umgeladen werden. Dieser bewegliche Teil lässt sich wie eine gewöhnliche Kiste von Haus zu Haus befördern und sogar bis in die Werkstatt, wo die Güter an Ort

Opening issue of the journal of the Bureau International des Containers, Paris.

2   What Is a Container?

Only the simplest possibilities seem to fascinate the organizing mind.
—John Cage

In January 1934, the first edition of *Der Behälter*, the proprietary trilingual publication of the International Container Office in Paris, appeared. It began with an article by the Italian senator Silvio Crespi, the chairman of the Italian Automobile Association and the founding director of the Bureau International des Containers (BIC). The article had the character of a manifesto. Entitled "What Is a Container?," it drew a bold transit-technological line from antiquity to modernity uniting sea, air, and land transport. Crespi wrote the following:

In the broadest sense, the word "container" (*cadre* or *Behälter*) denotes everything which holds objects of any kind. Thus, for instance, boxes were used since remote antiquity as a means of transportation. Wild animals were also carried in such boxes for the Roman arenas....
The modern container ... is no normal crate, but rather the mobile box of the carriage, divided into subunits. In this sense, the container divides the carriage and the truck into two parts: first, a portion that we will call the secure portion, and which is formed by the wheels, the springs and the platform, and second, the box. This box forms the mobile portion, can be detached from the secure portion, and can be

transferred readily from one means of transport to another.... In this way, the container makes possible international house-to-house transport via multiple modes of transport. Reloading is kept to a minimum, and the goods within the container are not touched at all.... In order to clarify the nature of the modern container, the Italians call it the "cassa mobile" (mobile box).... This view of the mobile truck body is always on our mind in our advertising for the container....

The narrow boats of the Phoenicians, the light carts of the Egyptians, the plastered streets and the triremes of the Romans, national and international postal transport, steam engines, cars and airplanes are the externally visible features of great ages of progress and civilization. In its inherent limits, the international organization of container transit could open up a new chapter in the history of transit.

The *Cassa Mobile*

At the end of World War I, there were experiments throughout western Europe in combined transport and container traffic; congresses were held and research societies were convened. The background of this was the crisis of the railway companies and the massive success of the truck as a means of transport. Transit by road was more flexible spatially and temporally as well as in price. Starting a truck transport company required a minuscule investment, relative to the railroad business.

An abundance of small and medium-sized transport companies pushed their way into the market and presented considerable competition for the railway. This development took place not only in the European industrial nations but also in the United States. In Germany, the railway situation was further complicated by the fact that production levels decreased dramatically as a result of defeat in the war and the burden of reparations. Even large companies could no longer take for granted that they could send entire railcar loads, and consequently they had to look around for alternative transport possibilities.

Politicians, civil servants, and representatives of the various interest groups considered how they could intervene in the situation with regulations. A decisive point in this development was the World Automobile Congress in Rome in September 1928, where the ways in which railway and road transport could be integrated into a transport association were examined for the first time on the highest international political level. The transport container was presented as a solution, a box detached from the chassis, a *cassa mobile* [movable chest], which could be transported in combined transit on a railway platform car as well as on a truck trailer.

In the same year, led primarily by the carriage construction industry, a research center for container transport was established in Germany. In 1930, under the auspices of the International Chamber of Commerce, the International Container Competition was advertised. To make the final decision, the International Container Committee was organized under the chairmanship of Silvio Crespi and composed of members of the following associations: the International Chamber of Commerce, the Advisory Technical Committee of the League of Nations for Traffic and Transit, the International Union of Railways, the Permanent International Office of the Automobile Industry, the International Association of Recognized Automobile Clubs, the Central Council for International Tourism, the International Federation for Commercial Automobile Transport, and the International Office for the Normalization of the Automobile.

The multilateral efforts, combining economics and politics, culminated in the founding of the BIC in Paris on February 22, 1933. It was not a fortunate date in world history: three weeks earlier, the Nazis had taken power in Berlin, and within a few years, all signs were pointing to world war rather than world commerce. But for the time being, the European axis seemed to work once again. Although the developments that would prove to be more definitive in the history of the container were taking place at the same time in the United States, global container transport—like world time, world railway, and world measurement before it—found its regulatory center in Europe. To this day, all serial and identification numbers are distributed by the BIC in Paris.

Seatrain

Although the advocates of container transport in Europe never tired of emphasizing that the United States was several critical steps ahead in development and had already accumulated some experience with various container transport systems, advocates in the United States were also following the developments in Europe very closely. At the same time that in Europe the systematic investigation of the possibilities of container transport was beginning at the Automobile Congress in Rome in 1928, an intermodal transport system by the name of Seatrain was established on the East Coast of the United States, integrating diverse modes of transit.

This system involved loading railway cars onto ships from longitudinal tracks by way of special cranes and stowing them below deck in multiple levels. Not only did this (supposedly) inspire the container pioneer Malcom McLean's "revolutionary" idea 25 years later, it also allegedly prompted the editors of the widespread publication *Marine Engineering and Shipping Age* to coin the term *container ship*. In honor of the commissioning of

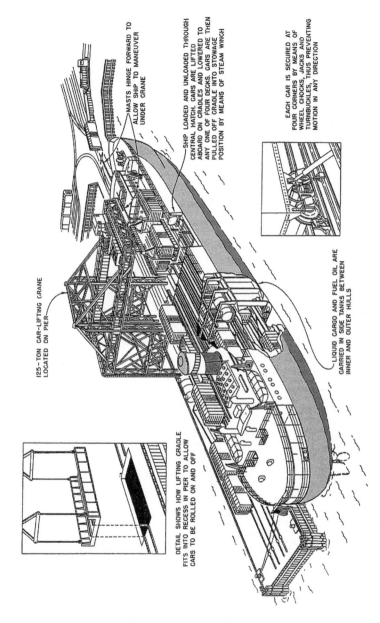

"How Seatrain Stows Railcars on Ships," *Port of New York* magazine, 1932.

two new ships in the fall of 1932, *Seatrain New York* and *Seatrain Havana*, which were specially constructed for the intermodal transport of railway cars—"the ship built around the cargo"—an article in the October issue entitled "Container Ships" stated the following:

> It is inevitable that the future will bring to the shipping industry newer and more efficient methods of transportation.... The conception of a new marine transportation system is one in which shipping serves as a link in a complete scheme of transportation, including rail, truck, and ship, providing door-to-door service. A move in that direction, which possibly indicates a trend towards a new phase of marine transportation and one which in effect extends the service of land transportation facilities, is seen in the activities of the Seatrain Lines, Inc., which will place the new steamships *Seatrain New York* and *Seatrain Havana* in service this month. In the case of these vessels, the process of design was reversed, the type of cargo, namely, freight cars, was considered as the starting point, with the ship built around the cargo....
>
> The type of vessel used by the Seatrain Lines, Inc., is a step in the direction of conforming ship transportation to the best and most economical practices on land. However, within the last few years, the motor truck, with its greater flexibility, has become a powerful factor in the transportation problem, necessitating the development of a standard container which can be used both by the truck and by the railroad as a unit of cargo transportation.

As in Europe, the consideration of a container transport system in the United States received its decisive impulse through the spread of trucks and their greater flexibility. The "Container Ships" article continued:

> While the future must decide the most economical type of cargo ship, present developments indicate that the trend of efficient shipping lies in the direction of conforming to the standard unit of shipment, the container. The Seatrain project uses the box car as a unit, but, if carried to its ultimate conclusion, a similar type of ship could be designed using the container unit and saving the weight of box-car underpinning and rails.

In 1936, an article appeared in the venerable journal *Annals of the American Academy of Political and Social Science* entitled "The Freight Container as a Contribution to Efficiency in Transportation." In this article, containers were defined, very much as they were in the article about Seatrain, as "of such size, that they can be carried with loads equivalent to a truckload or a portion of a truckload, on freight cars, highway trucks, trailers, or semitrailers and in or on water-borne vessels."[1]

Le Container

As the BIC resumed its public relations work in the 1950s after interruption by World War II, it propagated a systematic definition of container transport through exhibitions and brochures that stands to this day:

CONTAINER

Means of transport (box, removable tank, or similar transport vessel), that

a. Is of durable construction and resilient enough to be used repeatedly;
b. Is especially constructed to ease the transport of goods through one or several modes of transport without repacking the cargo;
c. Is equipped for easy handling, particularly when transferring from one mode of transport to another;
d. Is built such that it can be loaded and unloaded easily....

The term "container" includes neither vehicles nor ordinary packing materials.

Point c indicates that it is necessary to mechanize and, where appropriate, to automate the entire process; this leads to the need (not mentioned here) to standardize the elements of the system, which the BIC played an important role in as an umbrella organization. The last line, which demarcates the container from what it is not, makes clear what was of interest at that time: clarifying the conceptual change brought about by the introduction of the container.

Le Container is neither a vehicle nor ordinary packing material. Rather, it is a *means* of transport, a medium between transport vehicle and cargo that enables us to conceive of the transport process independently of its concrete form—structure, length, duration—and its concrete content. The container interrupts the interruptions caused by the transfer process between modes of transit by placing itself in and between them. In the words of the French philosopher and media theorist Michel Serres, it is a parasite that grafts itself onto the available traffic channels

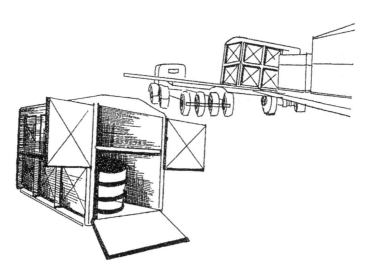

Model for intermodal large container transport from rail to road, *rationelle transport*, 1962.

and, in so doing, changes the entire system. With the spread of the standardized, intermodal transport medium of the container (and its "little sister," the pallet), an organization of transport is established that unites the various modes of transport on land and water into a chain.

Reentry

On an early morning in May 1966, the unloading of MS *Fairland* began in the international port of Bremen, Germany. Like balance beams connected to a crossbar at their outer ends, the linkage of the ship's own two gantry cranes jutted out horizontally over the railing of the ship. With these, the containers on board glided onto trucks that stood ready between quay sheds 16 and 18, in an open space prepared for just this purpose. Steel cables united the bridge running back and forth on the gantry crane with the spreader, a special gripping and holding device from which the containers hung and that held them in position.

The U.S. shipping company Sea-Land's MS *Fairland* was built exclusively for container transport; it could hold 255 35-foot containers. As such, it represented the first true container liner service between the United States and Europe; Bremen was the only harbor in Europe where it docked, apart from Rotterdam, Netherlands. But only a few people witnessed the event that is now seen as a historic date.

Despite the decades of precedent in land transit, no one in the port industry had experience with the large steel overseas containers and the loading system specially devised for them. Before the Second World War, individual wood or metal transport boxes were loaded onto ships. But these were much smaller and consequently easier to handle.

Unloading of MS *Fairland* in Bremen harbor, May 6, 1966. Bremer Lagerhaus-Gesellschaft.

The retired harbor master Bodo Meyer, who attended the unloading of MS *Fairland*, recounted the following in the 2006 documentary film *Die Container Story*: "I had no idea what a container was, really. And they told me ...they are large tin boxes, 35 feet long, they're unloaded with some kind of special device and reloaded.... Of course, you also had to just see that for yourself, ... that was new territory for us."

With the second box, an accident nearly happened: the container broke free from two of the four attachments and crashed down on the chassis of the waiting truck, caving in a portion of the driver's cab. Indeed, the driver had to be taken to the hospital, although he was, fortunately, not badly injured. The incident was not a bad omen, for the rest of the unloading went smoothly. The transport of goods by way of standardized containers and specially produced cargo handling equipment proved to be efficient and safe.

A year and a half later, the 40,000th shipping container was processed in Bremen. It mainly had to do with U.S. troop transports. In Hamburg, where the opponents and skeptics argued against the introduction of the new transport system longer than the skeptics in Bremen did, the arrival and unloading of the container ship *American Lancer* on May 31, 1968, in the Burchard wharf, reconstructed for this purpose and outfitted with a shore-based bridge crane, signaled the beginning of container transport.

The American Challenge

The development of container transport took place in parallel and reciprocally on both sides of the Atlantic, and its decisive impetus was land transit. Nevertheless, the large steel containers from the United States landing in Europe's ports after 1966 were regarded, and are regarded to this day, as objects of sea transport and as an American invention. They were part of the "American challenge," as the title of a highly regarded 1967 book by French journalist Jean-Jacques Servan-Schreiber reads. His arguments were also used in the transport and shipping industry to reinforce the view that a more comprehensive structural change was unavoidable, particularly if one wanted to find a European answer to the American container system.

Thus, it was no longer a matter of *whether* one could shift transport to standard containers, as was the case in the years before the war, but rather primarily of *how* it could best be accomplished. At least in the eyes of its advocates, containerization was coming like a force of nature. One of the first journalists in the German-speaking realm to offer answers to the practical questions associated with it was the economist and transport expert Walter Meyercordt. He dealt with the "economic meaning of container transport" as early as the 1950s, and he wrote and published a series of relevant texts. In 1974, he released a handbook for practitioners, the *Container Primer*. In the introduction, he once again held steadfastly to the relationship between the *Behälter* and the container:

Whereas a container has to do with the specifically American term for a transport container in general, the term "container" is nevertheless always associated with maritime transport. In all of their measurements, the ISO-containers are based on American constructions. As there was a move in the direction of the "container" around ten years ago, this explanation should be put first.... Because the container, the *Behälter*, is nothing more than a medium of combined transport: and combined transport has been the goal for nearly 150 years.[2]

ISO stands for International Standards Organization, the world organization for the establishment of international norms. In fact, the fundamental details of container transport—dimensions, materials, maximum weights, technical details of the handling process, cranes, and so on—were already determined in 1961 by ISO's Technical Committee 104 even before the subject of negotiation, the ISO container, left the United States for the first time.

The talk of a basis in American construction belies the fact that representatives from all industrial nations took part, sometimes quite significantly, in the negotiations of the new standards. For instance, the Hamburg port operator and containerization pioneer Kurt Eckelmann, the head of the ISO committee responsible for container dimensions, successfully moved to adopt the standards measurements agreed upon in 1961 by the American Standards Association. He did this because the Europeans were entirely unable to formulate a unified position; each nation involved made its own demands.

At a summit in Paris in 1964, ISO adopted the compromise of the Eckelmann committee and made it the basis of the unified dimensions for shipping containers agreed upon in October of the same year. Ultimately, the influence of a European was responsible for the international container dimensions of the American system—which are measured in feet (and elicit much head scratching in Europe to this day) because Europeans were far from able to speak with one voice at the time of the negotiations.

The distinction between *Behälter* and *container*, along with the term *Behälterverkehr* ("container transit"), which "always associates [the container] with maritime transport," is largely forgotten in Germany.[3] Yet Meyercordt fostered another misconception

(a) 20' All-steel container

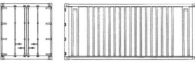

(b) 20' Steel-frame plywood container

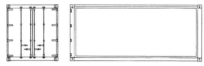

(c) 20' All steel open top container

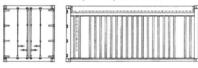

(d) 20' Flat rack

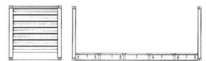

(a) All steel container.
(b) Steel-frame plywood container.
(c) All steel open top container).
(d) Flat rack.

What Is a Container?

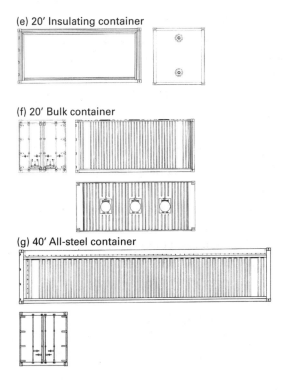

(e) 20-ft. insulating container—Max gross weight: 20,320 kg. Tare weight: 2,300 kg. Payload: 18,020 kg. Capacity: 26.7m^3. For chilled and frozen cargo without its own cooling device. On the front face (across from the doors) are openings for the cooling connections in the ship.

(f) 20-ft. bulk cargo container—Max gross weight: 24,000 kg. Tare weight: 2,350 kg. Payload: 21,650 kg. Capacity: 32.0m^3. For bulk goods such as grain, ore, coal, minerals, etc. Two valves on the door for unloading.

(g) 40-ft. all steel container—Max gross weight: 30,480 kg. Tare weight: 3,420 kg. Payload: 27,060 kg. Capacity: 67.0m^3. Breadth and height same as 20' container. For general lightweight goods.) ISO-container types. From: Friedrich Böer, *Alles über ein Containerschiff* (Herford, Germany: Koehler, 1984), 70, 71.

about the view that the container was an American invention. He only meant, however, that when something came to port in a European harbor in 1974 that was called a container and that adhered to ISO norms, it must have come from overseas.

The distinction, according to the criteria *Behälter*-Land-Europe (the old way) and Container-Sea-America (the new way), is an impermissible abbreviation from the perspective of 1970. It is much more sensible to speak of a competition of (intra)continental systems. Container transport is part of a redefinition of the relationship between land and sea that began long before containerization and that has both a European and an overseas (e.g., American), and a global component. Indeed, the container system is an important part of the history of globalization, but it is determined and continually redefined by many local relationships and by a network of diverse agents in various places, in distinct contexts, and in various functions.

Without minimizing the tremendous importance of the United States in the last 100 years, it can be said that what was and is considered American was never solely in the power of Americans to define. Americanism—a specifically European and, especially in Germany, a very widespread form of collective hysteria—played a catalytic role in the 1920s in the social disputes over all sorts of modernizations, and this has continued in the post–Second World War period.

Everything that was faster, more efficient, more modern, and more progressive was identified with America. There seems to be a nationalist—or, more accurately, a continentalist— tendency in the historiography of the distinction between (European) land transport containers and (American) sea transport containers that binds the rise of the container system with the grand tale of the rise of the United States as dominant world power (and the desired grand counternarrative of the "United States of Europe").

The American chronicles of the container system in particular tend to neglect both the European origins as well as the constitutive participation of various European and Asian agents in the process of containerization. They certainly do not show how local, national, and supranational interests of actors from various political systems and economic undertakings coalesced into a complicated mesh in a long process, with a degree of cross-linkage that can certainly serve as a model for the structure of the process of globalization in the second half of the twentieth century. Thus, the paradoxical situation arises that a history that claims to have as its subject a truly global system cannot free itself of the nationally and culturally delimited slant of its authors. And so the container becomes an American thing from the maritime world.

A General History of the Container

Even in the United States, the motherland of this specific form of new, large transport containers, confusion reigned at first. "Containers were the talk of the transportation world by the late 1950s," writes economic historian Marc Levinson. "But 'container' meant very different things to different people."[4] A basic reason for this confusion may have been that the word *container* in English—like the word *Behälter* in German—is a general expression for all manner of objects that can enclose something. However, because the container has become the core of a novel, complex, mechanical transport system, and as such is discussed worldwide, the expression suddenly stands for a very specialized object, defined in almost every detail—without having lost its potency as a general term.

Thus, along with the modern reinvention of the concept, its terminology has also had a notable trajectory. It has not only superseded the local, general terms for containing and transporting objects around the globalized world to become the undisputed name for the universal transport container. Rather, a proliferation of terms has occurred parallel to the spread of the transport system. *Behälter,* or containers, move more and more into everyday life and thus move increasingly out of the actual sphere of transport. All the daily necessities, everything that has formed culture since the prehistory of humanity, is now the content of containers, from food to waste. All of this has become the subject (or object, depending on which side one views things from) of containers in recent decades, from food containers to garbage containers. What was only recently still called a *bowl, barrel, bucket, carton,* or *car,* or possibly even a *bag, pouch,* or *pack,* became a *container.*

Container also increasingly pops up in metaphorical meanings. In relation to computers, it serves on the one hand as the designation of a certain type of program that operates like a transport container, and on the other hand as a spatial organizational structure for information. In other fields the term is used as a (defamatory) designation for a certain kind of architecture or a certain kind of space; as a (mostly derogatory) term for a certain philosophical, sociological, or physical understanding of space; as a designation for a new kind of room, perhaps comparable to the traditional *chamber;* or as a conceptual, spatially conceived bracket around a majority of things. In short, it is a material or imaginary unit of space that has the advantage of being easily filled with anything (and emptied again).

What, then, is a container? Even a quick overview demonstrates that the answer is not as obvious as it seems at first glance. The historiography of the container system is spread across diverse areas: the history of technology, the history of sea travel and of ports, the history of land travel, economic history, the history of communication and its technical media, military history, institutional history, and the history of hygiene.

Added to these are practical disciplines that employ the (conceptual) container as an organizing element, such as business administration, operations research, architecture, communications and information, design, and the supreme discipline of the container (so to speak): logistics. Sociology, physics, linguistics, or media studies could be mentioned as further areas of container questions. And not least, there is philosophy—although it was passé, in the view of many of its (not insignificant) representatives, by the time that (and precisely because) physical, conceptual, and metaphorical containers began to rule everything.

Therefore, the attempt to write a history of the container and an analysis of its present relevance cannot limit itself to the isolation and representation of the components of a single principle whose effects could be detected in all parts of the world: all disciplines, forms of thought, corporate divisions, and cultures, regardless of the concrete conditions. Indeed, this seems particularly obvious in the case of the standardized transport container, but is it (constitutively) self-identical everywhere—at least, the idea of it?

Thus, there are uncertainties, contingencies, and alternate possibilities for development in the rise of the modern standard transport container system. Moreover, there was never a single system, only continually competing technical solutions. In light of these factors, along with the spread of the term and its conceptual elements as a metaphor in so many different areas of knowledge and action, the intermodal container transport system seems less a unified technical solution than a modern ideal. One might call it a "steel utopia," in reference to sociologist Max Weber's famous formulation of capitalism as an "iron cage"—the incarnation of a modern myth of efficiency that radiates across many areas of society.

As a central medium in the spread and safeguarding of the rule of globalization—understood here as the global proliferation of the more or less free exchange of consumer goods under capitalist conditions—the container system reaches nearly every corner of Earth. However, it does not have blanket coverage but rather forms zones and corridors of its own temporal-spatial organization (as tight as its network may be). At its edges, wholly different principles can assert themselves, which is not to deny that the (standardized transport) container shows a tendency to expand its zones and subject whole continents to its clocked machine movement.

Nevertheless, if a container is moved out of the central realm of the logistical system, its ideal handling rules also no longer apply, or at least they apply only in limited form. An ISO container that is repurposed as a house remains a container, yet it becomes something else at the same time. A container can be punctured, sawed, stacked improperly, or placed upright as a sculpture. A container can be an imaginary conceptual unit without spatial expansion—indeed, temporary. A container can have rounded corners and slanted edges, be made of steel, be serially produced and compatible with conventional means of transport, but be neither stackable nor in accordance with international standards. In other words, and to emphasize once more, a *container* can be many things.

Container sculpture by Luc Deleu. © 2014 Artists Rights Society (ARS), New York / SABAM, Brussels.

Architecture—Other Words and Uses for *Container*

If one does an Internet search for the phrase "What is a container?" in German (Was ist ein Container?), one may find the website of the Berlin company containerbausysteme.de (http://www.containerbausysteme.de) that offers "insulated containers for the accommodation of people," with a relatively comprehensive attempt to limit the advertised object. Since the definition given on the first page of the site—"a rectangular or square cube for the housing of people or materials"—fails to confer the desired clarity, the following two lists serve to render concrete the advertised product. First comes a list that expands the realm of the intended object by all those types of space or technical artifacts that can also be attributed to the genus of the container, although they bear other names:

Other Words for Container

Container systems
Container modules
Mobile rooms
Mobile spatial systems
Mobile systems
Modular spaces
System container
System construction
Space modules
Space systems/mobile space systems
Space cells
Steel modules

The combinatorial game with a few words and the demure charm of concrete poetry shows clearly how the terms *container*, *module*, *space*, and *cell* become nearly interchangeable when they appear as components of technical systems to prepare standardized spaces and are granted the extensions *system* and *mobile*. Perhaps because this attempt at clarification also fails amid the systematic representation of the intrinsic components of the technical system, there is another list on the company's website that expands the view to the social dimension and lists the concrete uses:

What Is a Container?

Applications for Containers and Office Modules

Doctor's office

Construction management office

Contractor's office

Savings or banking container

Office building

Roof addition

Indoor addition

Hotel or hostel

Information box or information container

Youth recreation home

Ticket counter

Kindergarten

Hospital, bed station, medical building

Laboratory container

Exhibition container

Gatekeeper container

Clean room

School, school container, school pavilion

First-aid station

Sports building

Social space

Technology container

Administration building

Workshop container

Since the container superseded the barracks constructed on-site from prefabricated wood or metal pieces as the universal instrument for the quick preparation of functional spaces beginning in the 1970s, it has ruled this subarchitectural field of "temporary" use almost unchecked—whereby the common term *temporary* merits the relativizing use of quotation marks in two respects.

First, every use of something is temporary, so the formulation is ultimately tautological. Second, the uses legitimized as "temporary" regularly demonstrate that they are of remarkable endurance. From a home for asylum seekers to an equipment manager's shed, from a bank branch to a university, from an art association treasury to a guardhouse, an economy of fast spatial deployment has established itself in the most varied social fields and areas of function, the maxims of which are (or should be) mobility, speed, availability, multifunctionality, and cost effectiveness.

"Quick, simple, practical, cheap—four imperatives of all mass production in the name of progress," explain Axel Doßmann, Jan Wenzel, and Kai Wenzel, who deal collectively with the history of "portable buildings" in Germany from the 19th to the second half of the twentieth century in a study well worth reading.[5] These imperatives also apply to the industrial production of space.

The lists of the container renters do not only present a comprehensive inventory of the modern uses of containers or of the uses of their derivatives and structurally similar technical solutions. At the same time, they also reflect terminological and systematic developments from the idea of a mobile room cell to a living unit made from used standardized shipping containers. Historically, the concept of the container appears at roughly the same time in architecture as it does in the transport network—namely, in the 1920s. Because of rapid population growth, urbanization, technologization, and war, the unity of the "anthropotope," to use a term from philosopher Peter Sloterdijk, of human housing in its evolved, (premodern) technical and social-functional contexts, is destroyed.[6]

What Is a Container?

Containers in the service of society: (a) police container, Hamburg, 2002; (b) Waldorf School, Berlin, 2004; (c) security checkpoint, Munich, 2006; (d) Information Center, Berlin Holocaust Memorial, 2004; (e) first-aid station, Munich Oktoberfest, 2005. Photos by Alexander Klose.

This circumstance forms the historical background to the rise of the container principle in architecture, which architects such as Walter Gropius and Le Corbusier, to name only the two most famous, began to formulate at the beginning of the 1920s. These plans and programs were implemented on a larger scale only after World War II—comparable to the development of container transport. It is important to see that the grievance brought to this day against *box*, *cell*, and *container architecture* does not merely represent a metaphorical sharpening. Rather, the defamatory speech is aimed directly at the core of modern building.

Container Space

Beginning in the 1950s and strengthened in the wake of the spatial turn since the 1990s—that is, the return of social and humanistic theories to spatial matters—social-scientific theories have been developed that are meant, on the one hand, to counteract the forgottenness of space in the classical conceptions of society, economy, and communication. On the other hand, they depart explicitly from the notion of an objectively given, empty "container space" in favor of a model of spaces determined through social interaction. Rather than proceeding from an understanding that assumes space to always be equally present across all times and places (causing it to disappear as its own quantity), these theories assume that the respective spatial *perception* is determined by countless temporalities, travel speeds, and forms and ranges of social interactions.

As with all (modern) concepts of spatial models, these new positions also received their inspiration from physics—namely, from the critique formulated since the beginning of the twentieth century of Isaac Newton's notion of an empty and immutable geometric space. Albert Einstein voiced this criticism in a 1953 essay on the history of physical perceptions of space with the influential formulation of the "box space":

> Into a certain box we can place a definite number of grains of rice or of cherries, etc. It is here a question of a property of the material object "box," which property must be considered "real" in the same sense as the box itself. One can call this property the "space" of the box. There may be other boxes which in this sense have an equally large "space." This concept "space" thus achieves a meaning which is freed from any connection with a particular material object. In this way by a natural extension of "box space" one can arrive at the concept of an independent (absolute) space, unlimited in extent, in which all material objects are contained. Then a material object not situated in this space is simply inconceivable; on the other hand, in the framework of this concept formation it is quite conceivable that an empty space may exist.[7]

Einstein's characterization of the Newtonian notion of space, which he refutes with his theory, culminates in the formulation of space as the container of all material objects. Thus, at precisely the same time that the concept of the universal transport container began to take on concrete, material form, the famous physicist, eminent in all fields of knowledge in his day, introduced the term *container space* into the discourse on conceptions of space (a concept that is still widespread today, despite all the criticism directed against it, since it is derived from an intuitive perceptual praxis developed over centuries).

Since then, the *container* terms have evolved in diametrically opposed directions. Whereas the transport container is brought into a system that is technically differentiated and standardized and is being more and more spatially and materially determined in all its parts, the notion of the container in connection with Einstein's theoretical frame expands to arbitrary dimensions and global applicability. This reaches from the concept of the body and of human beings as containers of (changing) subjective definitions to the concept of society as a nationally and territorially defined container of language, culture, and political organization and to the concept of the planetary realm, or space, as a container of certain physical qualities.

What binds logistical, physical, and cultural container theories together is their constitutive emptiness, or evacuability, which makes the container a universal receptacle. It stands in the center of most container theories, whether negative or positive. Likewise, the philosopher of art Hannes Böhringer writes the following in a furious essay on the container:

> The container holds everything, but all that is shit: separated, distinguished and yet in its isolation made uniform with all the others. The content-holder, the steel box, in which being (*Sein*) crumbles into beings (*Seiendes*) (Weber-Heidegger), is a toilet, drain and container, devaluer of values (Nietzsche), closed space that is fully empty. The container is no longer a vessel that captures the world and embodies man, it is the magic black box of nothing, end of man and of the world, the death of God, the end of the end and of the beginning, endlessness of the tentative, a vacuous container that loses what it contains.[8]

The Subject Container

The linguist George Lakoff and the scientific theorist Mark Johnson show in their book *Metaphors We Live By* how the conceptual division between subjects and objects inherent to language fundamentally structures human perception and leads to the use of "ontological metaphors"—that is, "ways of viewing events, activities, emotions, ideas, etc., as entities and substances."

According to Lakoff and Johnson, the primary subspecies in this genus of the ontological metaphor is a container metaphor: "Each of us is a container, with a bounding surface and an in-out orientation. We project our own in-out orientation onto other physical objects that are bounded by surfaces. Thus we also view them as containers with an inside and an outside." As a rule, ontological metaphors are not perceived as such, but rather as "natural language" that names the perception of things directly: "self-evident, direct descriptions of mental phenomena."[9]

According to this theory, by projecting their container-like self-perception onto the objects around them, the subjects separate themselves as closed-off, container-like entities from the containerlike objects in their environment. They constitute themselves as subjects separated by borders from the objects that surround them. Thus, humans conceive of themselves as containers through the metaphors that they choose within their "concept system," and they likewise form their surrounding (material) beings and objects. This is a variation of the modeling of things, people, and spatial situations as containers sketched out in the last section.

In contrast to the mostly negative connotations of the container terms in sociology, the container-contained theory of psychoanalyst Wilfred Bion refers essentially positively and productively to the image of the subject as a container to be filled and emptied. In this theory, a fundamental aspect of the therapeutic relationship to patients with severe personality disorders is "projective identification." It consists, in essence, of a patient projecting part of him- or herself, an affect that is experienced as unbearable, onto an external object: the therapist.

The therapist serves as a container—that is, as a temporary storage for the fractured and projected contents and their associated feelings: as a rule, "archaic rage processes." In the course of therapy, the therapist gives these contents back to the patient. They are not unchanged—and this is the highlight of the matter—but rather have been "predigested" or "reflected" by the therapist-container.[10]

In contrast to the majority of container concepts and metaphors, and particularly in contrast to the transport-logistical container system, the therapist-containers are anything but indifferent about their contents. If the therapy is effective, the patient can better accept the fractured portion of him- or herself, whether identifying with it or distancing from it.

The notion of the container is similarly active in Melanie Klein's theory, to which Bion refers. Melanie Klein was a student of Sigmund Freud's who worked extensively with the problem of early childhood trauma. In her psychoanalytic model, the mother contributes decisively to development as a container for the projected feelings of her child, which she returns with evaluations.

On the various levels of the transmission or identification process, a symbolic transformation takes place in the form of a nesting of the container and the contained: the object that serves as a container for unformulated (or unformulatable) feelings on the first level becomes itself contained in the container of the individual on the next level. Similarly, on the interpersonal level, any group can become the container for the statements of its members (for their feelings, thoughts, self-representations, and actions) and can absorb and contain them.

The Data Container

Matryoshkas, the colorfully painted wooden nesting Russian dolls, are among the most famous Russian exports. They serve as an organizational model for one of the most successful container principles, the Matryoshka Principle, which has found its way into highly varied contexts. In business management, in production and logistics, and also in communications and information, it refers to models of recursive organization. An object or a model is recursive if it contains itself as a part, just as each Russian doll contains a smaller version of itself inside it.

In the definition of recursive functions, there is at least a call for the same function. They are also referred to as Matryoshka functions, for illustrative purposes. Such organizational models are spread out in (hierarchical) tree structures, such as the data system on personal computers, in which each folder may contain other folders. The model of the black box, which is central in cybernetics—the theory of control systems—and in systems theory, is also often part of a recursive schema. For example, when an entire situation is modeled as a system, the constellation within the black box is a subsystem within this system, which in turn contains subsystems whose regulation can be simplified and optimized by modeling them as black boxes.

These are certainly not the only conceptual, symbolic, or metaphorical containers in the world of data processing. In the intuitive conception or representation of computer programs, there is a mechanism that operates in a way very closely related to the formation of perceptual concepts through ontological and orientational metaphors described by Lakoff and Johnson. The boxes in communications and information are not containers in the physical sense. Rather, the programs and program components that are referred to as such are symbols arranged in lines; they are code.

Their functional definition and their location relative to other symbols make them into containers. Because their commands precede the other parts of a file in reading order and because computers, beginning with Turing's Universal Machine, execute commands symbol by symbol, line by line, and batch by batch, they can form something like the definitional framework. They can function, for instance, so that programs for which the content of a certain file is irrelevant can read only the framing lines of code, execute the corresponding commands, and transmit the entirety of the remaining code without processing it.

The notion of boxes or containers is not necessary for the functioning of these programs. Nevertheless, saying that only the code is real and that all these notions that one has about it are fictional would not only be too simple, it would be simply wrong. Computer users as well as programmers create schemes, mental models or shapes, in order to imagine organizational structures and ongoing processes. Such schemata are a fundamental part of intuitive models for the understanding of processes of computer programs, because paradigmatic models such as the IPO principle (input, process, output), which vastly simplify the complex program structures, play a central role in professional program development. The same is true for design patterns, like the distinction between *tools* and *materials* in office software.

This begs the question of whether ontological metaphors may not also be constitutive in the programming process, since notions of containers and their historical substantiation are within certain modern modes of transit, such as omnibuses or standard containers. In other words, could a program that serves to bring about the *transport* of certain data *contents* like films or music at the end of the twentieth century have been conceived of as anything but a *container*?

The artists' group etoy is making the homology between goods transport and data transport productive in a particularly catchy way. Etoy, which became famous as a group of pioneering

Internet artists in the mid-1990s, has worked with ISO containers as a central element of its art since 1998. The steel boxes that etoy calls *tanks*, painted in its corporate identity orange and furnished with its logo, are part of its artistic strategy, of a deceptively genuine imitation of the aesthetics and practice of commercial enterprises that the artists themselves refer to as *corporate sculpture* (inspired by the groundbreaking German conceptual artist Joseph Beuys, who triggered the concept of "social sculpture"), in which the form of the internationally active business is sculpture.

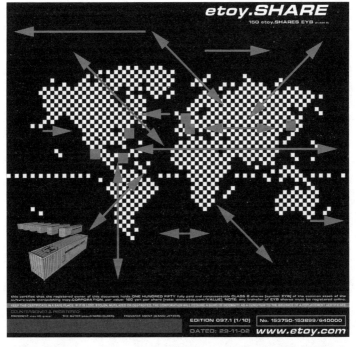

Artists' group etoy's container distribution diagram. © etoy. SHARE-CERTIFICATE No. 97.

As such, the boxes serve as symbols as well as mobile space for artistic actions that take place across the globe, mostly in cities, and part of these actions is the delivery and installation of the containers themselves. In a strategy paper from 2004, etoy's members write about their containers and the relationship between goods and data transport:

> The multifunctional etoy.TANKS are a crucial part of the etoy.GESAMT-KUNSTWERK: studios, corporate sculptures and walk-in-webservers ..., modular office bricks that travel the physical world in the same way data packages travel the internet: every etoy.TANK is a TCP/IP-PACKAGE traveling through space and time according a very elaborate global transmission protocol to distribute etoy.CONTENT.... The complementary aspect and the melting of global traveling, trade and complex logistics with experimental digital tools and services are in the artistic focus of the etoy.CORPORATION: etoy.AGENTS are exploring the space between the digital and the physical.[11]

Icons of Globalization

What does it mean when a theater festival illustrates its advertising poster and postcards with stacks of containers, as the 6th Politics in Independent Theater Festival (*Festival Politik im Freien Theater Berlin*) in Berlin did in 2005? Why would an art exhibit move its entryway and its service booths, information, bookstore, and coat check into containers and spread them across the entire city, as with Documenta 11 in Kassel in 2002? Or why would it just present all its artwork in containers, as with the art project of the Academy of Fine Arts, Munich, at the Federal Horticulture Show in Munich in 2005 that took its name and its logo, *Evergreen*, from one of the largest container transport companies? What does it mean when a film has its hero look for shelter and then find himself in a discarded and repurposed shipping container, as in Aki Kaurismäki's *The Man without a Past* (2002)?

What Is a Container?

"Desire." Postcard for the 6th Politics in Independent Theater Festival (Festival Politik im Freien Theater), Berlin 2005.

In each of these cases, the presence of the transport box points to the loss of firm ground in formerly safe conditions through the incursion of the world. Capitalism and globalization sweep away older, metaphysically grounded order by establishing a new time-space order in which Earthly relations are spread across far greater distances than ever before. Their powers are the effects of secular networks, even if they often push through as brusquely as events whose causes were previously found in the heavens and attributed to the gods, or at least to destiny. Traditionally, theater, movies, and art exhibits function as thresholds at the boundary between the small order of the human world and the grand order of the divine—where cultural production itself could stand metonymically. Today there is the container. It embodies a new order of global immanence, life "in the inner world space of capital" (to cite the title of a book by Peter Sloterdijk), where (supposedly) transcendental powers can be found only in the opaque workings of bureaucracies and technologies and their accumulation effects.

The container, itself a significant agent of this development, has become an icon of globalization. To be precise, it is not the standardized painted steel box itself, but rather its image, that is reproduced 10,000-fold in various media: from theater to television, from an avant-garde art exhibit in a world metropolis to the local section of a newspaper in Timbuktu, from documentaries critical of globalization to Hollywood blockbusters. But the iconic status of the steel box has made the analysis of its role in the process of globalization more difficult. Its specious evidence as a superior means of transport and a defining medium of globalization occupies the space of all possible answers and consequently obstructs all questions that appeal for more complex answers.

The literary scholar Uwe Pörksen, who has dealt with the function and mechanisms of such iconic images in modern social discourse, suggests the term *visiotype*—analogous to the term *stereotype*. These images are "more important than the catchphrases" and are "key stimuli of consciousness."[12] Today the transport container has become just such a key image, a global visiotype that professes to make further explanations superfluous.

The success of the metacontainer has brought about a metareality in which containers and globalization have always formed a firm and fast tautological unit—a reality in which the container revolutionized the global economy from the moment of its appearance; a reality in which container technology moved steadily and unstoppably forward to bring "the principle of globalization" to dominance, as it was called in the title of a Swiss art magazine in 2003.[13]

This metareality consists of a bastion of belief in progress and the apotheosis of rationality, regardless of whether this process is interpreted as philanthropic or branded as misanthropic. The reality is based on a mythical foundation that attributes technical and social development to ominous powers of the economy and market.

In the container world, the global economy appears as a model kit or toy landscape of colorful building blocks. The similarity between containers and Lego blocks is striking. The taxing work of hard men who transferred the goods was still in the public view in 1950—think of the portrayal of dock work in Elia Kazan's impressive 1954 film *On the Waterfront*, with Marlon Brando in the lead role. It became a nearly floating action of almost childlike lightness through standardization and mechanization, a "container ballet"—or so was the impression it gave.

It is fitting, then, that the steel containers, as big as houses and weighing a ton, moved only by powerful machines, are preferably spoken of in trivializing metaphors. They are termed *boxes*, *cases*, *bins*, *cartons*, and *cans*. In German stevedore jargon, the loading and unloading of container ships is called "spinning boxes," as though the boxes had never left the reach of human hands.

"Where the world becomes picture," philosopher Martin Heidegger wrote, society's view of itself becomes a view of a building or circuit diagram. "Understood in an essential way, 'world picture' does not mean 'picture of the world' but, rather, the world grasped as picture."[14] In the world picture that the transport networks of goods and information jointly create, the physical world is almost disembodied. If the notion of the Internet as an

Model or reality? Work or play? [World trade. Moved by Linde] Advertisement from Transport Manufacturer Linde. Linde AG.

immaterial communications system succeeds in concealing tens of thousands of tons of cables, circuit boards, and housings, then the vision of the container world succeeds in making millions of tons of heavy metal appear weightless and frictionless, moved as if by magic. And it is, above all, this image of a frictionless, well-ordered organization through standardized containers that makes containers so attractive as a structural metaphor. It evokes the image of a neutral medium, a pure movement of units of information, production, and consumption on the circuits of systems.

Box Landscapes

In the 1920s, attempts were made in a series of exhibitions to make sociological, medical, and economic advances accessible to a broader public by way of a three-dimensional model of curve diagrams. Contemporaries such as Walter Benjamin took note of these landscape representations of statistical data with some concern. But this new kind of popular treatment of information was well received by the public. Thus, it contributed in no small way to the popularization of demography as a political instrument. In the wake of the rise of infotainment, such "curve landscapes" have become so natural that even most children understand them. They have become normal pictures.

For a few years, "box landscapes," the arrangement and stacking of containers on a ship or in a harbor, have presented the image of an economy complementary to the curved landscape. "Trade Makes Container Waves" reads the caption beneath a photo of the overseas dock in Hamburg harbor in June 2006, which appropriately combines the maritime image of the economy with the new edgy medium, in the progressive newspaper the *Tageszeitung*. Like stock market graphs and their function, as examined by the economic historian Jakob Tanner, pictures of fully loaded container ships and busy container terminals act as "motivating and optimism-inducing icons."[15] The evidence of the container images is related to the productivity of enterprises and nations, to imports and exports that are no longer primarily presented in terms of money and weight but rather as money and standard transport units, or TEUs. Many boxes are good; a few boxes are not so good.

Whether containerization has actually simplified world goods transport or simply pushed the complexities to another level is a difficult and open question. In any event, containerization makes possible a new, extremely reduced pictorial representation that unites the level of abstraction of the quantitative expression of volume in numbers with the concrete materiality of the transport processes. Since the latter is always limited to the same forms, varying only in a narrow spectrum of formats, colors,

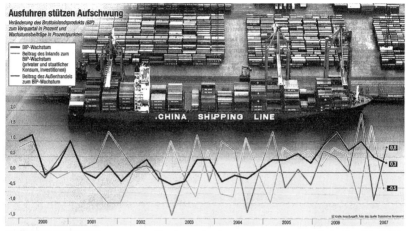

Exports underpin boom. Quarterly changes in gross domestic product (GDP) in percent and growth shares in percentage points. Combined box and curve landscape for the visualization of economic processes in the business section of a newspaper. SZ-Grafik © Ilona Burgarth, *Süddeutsche Zeitung*, 24.08.2007. Photo © ddp. *Source:* Statisches Bundesamt.

and labels, the visual depiction can concentrate on this symbol. There is no need to choose between numerous goods visible in the transport process, as in the time of classical cargo transport.

Transport space and administrative space, the cargo hold on a ship, and the stowage plan on paper or on the monitor are filled in the clearest way when containers are stacked on top of one another. Perhaps this is the most astonishing accomplishment of containerization: that it unites so directly the vast space of transport with the small space of the office. Although in the wake of containerization transport volumes have risen tremendously, this has made the process more open precisely because it removed the transported goods from view and from reach. In contrast to the clutter of bags, bales, barrels, and crates, large standardized containers are *countable*—not like traditional cargo shipping, in which an outsider could never know what kind of loading units were hidden in the belly of the ship.

The picturesque heaps of the various goods and packaging that shaped the iconography of the port until well into the second half of the twentieth century presented a problem that resolved itself with containerization, at least for information graphics. Since global goods transport reorganized itself, it has met the requirements of a plain manner of representation for pictorial statistics.

Containerization accomplishes what Otto Neurath, the pioneer of simplified graphical depiction of complex facts through pictograms, said in 1926: "One should represent a fivefold quantity with five figures, a tenfold with ten! That is, rows of small men, small wagons, small cars, etc.!"[16] It concentrates a complex technical and social process into *one* mediating and integrating thing that both executes and represents this process. The container is not only an icon but also a pictogram of globalization.

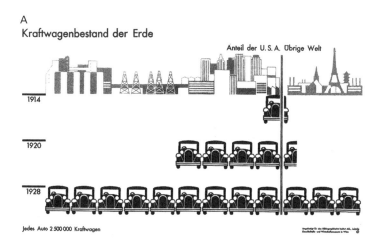

Countability as principle: (a) Global amount of cars (U.S. [left]/rest of the world). Otto Neurath's pictorial statistics, 1928; (b) model depiction of a container ship, 2005. © Hapag-Lloyd AC. Hamburg. Used with permission.

Viewed up close, the container system destroys every human benchmark for size. Try to find a point on the dock of a container port where a container ship can be seen with a glance. But the lofty size of the container industry equipment can be escaped by moving vertically. Every report on world trade usually takes place from such an elevated point of view. What was still reserved for a minority of observers in extreme situations at the time of Ernst Jünger's analysis of the worker-soldier (in his book *Der Arbeiter*, first published in 1932), the planetary perspective of a mobilized world is now an unquestioned basis of reality, individually controllable for every Internet user since Google Earth. Thus the media-technological capabilities are available to transform the vast mechanical network to a small image. This transformation, in turn, is the prerequisite for views of the container world that can simply generate very legible images or models and for container terminals and ships to function as pictorial generators of statistics.

The standardized mobile transport unit is a universal container and universal container name, a general metaphor for all things fillable and evacuable, a concept in the construction of space, a form of thought and order, a symbol, an icon, and a pictogram. Moving within this spectrum are the answers to the question "What is a container?" Silvio Crespi, the Roman senator, Automobile Association chairman, and pioneer of containerization, presumably had no real idea when he opened the Pandora's box of container meanings that the more success his principle was awarded, the more his questions would become unanswerable.

April 26, 1955: Birth of the large container system, loading of the *Ideal X*. © Sea-Land.

3 Sea-Land

I don't have vessels,
I have seagoing trucks!
—Malcom McLean

And I saw a new heaven and a new earth:
for the first heaven and the first earth were passed away;
and there was no more sea.
—The Revelation of John

Woe, when you feel homesick for the land ...—
and there is no longer any "land!"
—Friedrich Nietzsche

On April 26, 1956, on a pier in the New York–New Jersey harbor in Hoboken, the *Ideal X*, a ship of the Pan Atlantic Steamship Company, was loaded by using an unusual new process. The *Ideal X* was a rebuilt American T2-tanker from World War II. Dockworkers secured large aluminum transport boxes onto its deck, which was reinforced by a second layer of steel plates, perforated at regular intervals by rectangular holes. Jutting from the bottom of the boxes were four pins, which fit precisely into the holes provided in the ship's deck. The boxes were actually truck trailers, separated from the chassis. The dimensions—8 feet 6 inches high, 8 feet wide, and 33 feet long—represented a compromise between the road regulations of the American states crisscrossed by the trucks during land transport, the limits of the port, and the size of the stowage area on the deck of the tanker. Fifty-eight trailers were lifted by cranes directly onto the ship from the trucks parked on the dock.

When the *Ideal X* arrived in the harbor of Houston, Texas, five days later, the entire procedure repeated itself in reverse order. Fifty-eight of the same kind of trucks received the trailers and simply drove away to reach their destinations in the American South. The experiment with the truck trailers proved to be a success, and the enterprise expanded. In 1960, the Pan Atlantic Steamship Company was renamed Sea-Land Service Inc., "to better describe the services offered," the company's press release stated at the time. Thus, the fitting label used from the beginning for the new land-water transport service became the company's name.

More and more sea and land transport companies followed Sea-Land's example and sent out trailer ships (as they were called before the term *container ship* came into use). As a result, that date in April 1956 soon came to represent the birth of container transport. The story of the man behind this pioneering journey, Malcom McLean (1913–2001) from North Carolina, has been told so often since then that it has taken on the stuff of legend and fable. It is the story of the classic self-made man, a newcomer in an economic field constrained by tradition and law who persevered with unconventional ideas from a one-man trucking company to a captain of industry and, in the process, gave a decisive push toward the globalization of world commerce. There is hardly a text about the container system that does not contain at least a brief biography of McLean and his pioneering company, Sea-Land.

In the container's origin stories, technical elements from various sources converge: poor winter weather, cotton pressed into bales, archaic loading techniques, the value of native soil, businessmen unwilling to innovate, obstructionist bureaucracies, and a risk-taking banker. McLean himself, who was a master of creating reality in word and deed, also brought numerous container stories into circulation. Over the years, as both the legendary protagonist and the narrator, he contributed significantly to the fable.

The Container Story of Malcom McLean, a Farmer's Son

The fable of the "invention" of container transport goes like this. A young, clever, and strong-willed businessman turns the entire international transport industry upside down by implementing the simple yet powerful idea of loading large containers, instead of their small individual contents, between land and sea transport. The problems of cargo shipping were the high cost and the excessive idle periods in port. The problems of land transport were, on the one hand, bitter pricing wars resulting from overcapacity and, on the other hand, the lack of tariff and organizational flexibility because of official overregulation.

Without regard for prevailing conventions and definitions, McLean followed his vision, encountering dogged resistance from all sides and driven by a wholly unsentimental commercial rationality. His highest aim was to lower costs: "You know what freight is. You can look it up in the dictionary, but I'll tell you. It's something added to the cost of the product."[1] McLean's advance compelled authorities and competitors to rethink things. The moral of the story: Trust the rational view of things. Don't let yourself be slowed down by petty concerns or political or sentimental resistance. In essence, a good idea or a correct concept will always prevail in the end.

Embedded in this capitalist success story of the superiority of rational commercial calculation is the typically American legend of a man's rise from a humble background with the inspiration and implementation of a vision. Malcom McLean was a farmer's son from the American Southeast who had no money, but he had good ideas and a big heart, or so the legend might begin. Because the land that his ancestors had cultivated for generations was no longer productive, he began a career as a transporter of goods. He saw success, so he owned one truck and then eight, with drivers working for him. He called his business the McLean Trucking Corporation.

But luck was not always on his side. An unusually harsh winter with heavy ice and snow as well as the sudden collapse of the local textile industry, which provided his company's main transports into northeastern cities, nearly brought McLean to ruin between 1936 and 1938, about five years after his promising start. Debts resulting from the massive loss of contracts as well as damage and loss caused by accidents forced him to climb once again behind the steering wheel of his last remaining truck.

Thus it happened, on a morning in 1937 just before Thanksgiving, that he arrived in the harbor of Jersey City, New Jersey, with a load of cotton bales. He had driven all night to deliver the bales on time, only to find that the dockworkers were not prepared to unload his truck. He waited the whole day on the dock and watched as the workers loaded things by hand, box by box, barrel by barrel, bale by bale. Apart from utilizing the support of a few cranes, carts, and nets, this effort was powered by muscle,

Sea-Land

Dock work in 1950s New York was like that in the time of the Phoenicians. Film stills from *On the Waterfront*, 1954. © Horizon Pictures.

not very different from how Phoenician trade ships were loaded and unloaded on the Mediterranean some 3,000 years before.

Then, according to his own version of the events (published in the same 1994 magazine article), McLean had an inspiration: "Wouldn't it be great if my trailer could simply be lifted up and placed on the ship without its contents being touched? If you want to know, that's when the seed was planted."[2]

Over the years, several alternative versions of the origin legend have emerged. In one of them, McLean drew inspiration from observing the stacking principle of a cigarette vending machine. Another popular variation claims that McLean had been aggravated for years by the enormous losses in transporting beer from Germany. There is a kernel of truth to this story, for the first cost calculations for McLean's idea were undertaken in 1955 on the basis of McLean Trucking's beer transports from Newark, New Jersey, to Miami, Florida. The study's spectacular result was that beer transport in containers would be 94 percent less expensive than conventional cargo transport. Nevertheless, it would be another decade, with the beginning of transatlantic liner service in 1966, before German beer arrived in the United States in containers.

Initially, McLean's firm ran exclusively along the East Coast of the United States. On the return trip to the South, it carried textiles, shoes, shaving products, baked goods, and alcoholic beverages, among other things, and on the way back to the North, tobacco and cigarettes. After the hard times of the Depression, things were looking up for McLean Trucking.

At the beginning of the 1950s, during the height of its success, the company had 2,000 employees, more than 1,000 trucks, and 37 terminals in all the cities along the U.S. coast from Maine to Texas. McLean Trucking became one of the largest and most renowned trucking companies in the nation. That was when, according to one version of this tale, he reconsidered the container idea that he had conceived in the hard times before World War II and began to examine the conditions for its implementation, which culminated in the study of beer transport costs mentioned earlier.

The other version claims that he never lost sight of the idea and merely waited for the right time to put it into action. There are two distinct concepts of history hidden behind both versions. One might summarize the first as accentuating the "hero" aspect of the entrepreneur as a go-getter shaping his reality. In contrast, the second emphasizes the historical efficacy of ideas. A third possibility lies in highlighting the dynamics of technical systems and large structural contexts. The history of containerization vacillates among these three variations, according to whoever is telling it and with which interests.

In any case, as the legend continues, at the first available opportunity McLean acquired a small, run-down tanker shipping company named Pan Atlantic Steamship Corporation. He found no partner in the venerable shipping industry willing to get involved in the risk of trucking. In May 1954, he also bought the previous owner of Pan Atlantic, the Waterman Steamship Corporation, along with its freighters, docks, and wharfs. However, their value exceeded that of McLean's operation considerably.

In order to make the spectacular takeover a reality, he required an entirely new kind of financing, more than 80 percent backed by credit. The Waterman acquisition went down in economic history as the first leveraged buyout, a financing model based largely on outside capital, without which the wave of acquisitions since the 1980s would not have been possible. In the process, McLean's enterprise was turned into a public company.

Deeply in debt, the land transport firm McLean Industries emerged from the takeover, transformed into something entirely different and much larger. As the American cartel laws of the day forbade a company to operate by two distinct modes of transport, McLean threatened to upset the Interstate Commerce Commission, which was tasked with the regulation of transit. Consequently, McLean unceremoniously sold his share in the McLean Trucking Corporation, the enterprise that he made (and that made him) great.

Thus the trucker became the director of a shipping company—a shipping company that, however, did not aim to operate ships; instead, it would operate "seagoing trucks," as McLean put it in the famous witticism that is quoted at the beginning of this chapter. Despite taking out more loans, the new container transport company prospered and expanded in the years that followed.

It is notable that the legend of the visionary who consistently followed his inspiration, and the fable of the right idea that asserted itself, is complemented by yet another story: the value of native soil. This strengthens the land aspect of the container story in a surprising way and gives McLean's path from regionally rooted trucker to director of a globally active shipping and transport company an unexpected character. In an *American* magazine article, which was published at the height of his initial success as a trucker in 1950, McLean wrote about his values and his philosophy. His family descended from a "daring" Scottish farmer who had set out with his family for the New World several generations earlier. They remained farmers—"naturally," he said—in their new home, the "rich, fertile soil" of North Carolina.[3]

If one follows this depiction, then America was not a new *coast* for McLean's ancestors, behind which the unknown and unpredictable waited, be it destruction or promise. Rather, America was new *earth* for them, in which they would continue with what they had done in their home left behind: cultivation.

It was only because this new soil also ceased to be new after generations of extensive use that McLean looked around for other sources of livelihood. Although some of his classmates tried to convince him to go with them to New Orleans and be hired as a seaman, he decided to stay in North Carolina. First he stacked cans in a warehouse. Then he ran a gas station. Finally, he became a trucker. One of his first large jobs consisted of driving around mud and earth from Works Progress Administration construction sites, a public road-building program that was part of President Franklin D. Roosevelt's New Deal. His future success was possible, according to his retrospective interpretation, because he remained committed to his native soil.

Even though McLean's path led from can stacker to millionaire, his success differs from the American dream in one decisive aspect: Whereas the latter consists of setting out from humble beginnings and finding one's luck *elsewhere*, McLean acted according to the motto: "Opportunity begins at home." Had he been in his ancestors' place, he would presumably have stayed in Scotland. (As if to support this, he changed the spelling of his first name from *Malcolm* to the original Scottish version, *Malcom*, after he had already gone from national trucker to international container transporter.)

It seems that McLean's bond to his American homeland stands in irritating contrast to the uprooting dynamic of the flexibility and globalization of capitalism in the second half of the twentieth century that his company inspired. This yields a thoroughly contradictory (though supposedly not so rare) constellation: the innovative entrepreneur with a global horizon as homebound traditionalist and local patriot, the builder of global systems as profiteer of a local network, the "revolutionary" as social conservative, and the great fluidizer of transport as lover of the firm and grounded.

Before he entered the container ship business, McLean had never left the safety of firm land and set foot on the planks of a ship, as he himself emphasized. The story of the great pioneer of global container transit is largely a continental one, rooted in the soil of his home state in the southeastern United States. As we will see, this is not the only paradox produced by the modern revisions of the relationship between sea and land.

On the Drying of the Sea

On April 8, 1838, the *Great Western* departed from Bristol, England, for New York on its maiden voyage. It was the first steamship specially constructed for transatlantic passenger and parcel transit and the first that offered liner service. Its designer, the English engineer Isambard Kingdom Brunel, conceived of it as an "extension" of the Great Western Railway, an iron railroad being built under his direction at the time to unite London with southwest England and Wales.

As a continuation of the railway by other means, the steamship is the maritime equivalent of the transport medium that the nineteenth-century German poet and critic Heinrich Heine praised for killing space and leaving only time remaining—an often-cited expression that marked the opening of the railway from Paris to Rouen and Orléans in 1843. The new principle ended the zigzag tradition, thousands of years old, of sailing ship travel. The steamship took on not only the straight path of the continental rail lines but also the idea of an itinerary. With the

Great Western, the transatlantic passage, which lasted more than a month on average by sailing ship, could be reduced to 15 days (westward) or 14 days (eastward). A contemporary English journal, the *Quarterly Review*, said the following in 1839:

> We have seen the power of steam suddenly dry up the great Atlantic Ocean to less than half its breadth.... Our communication with India has received the same blessing. The Indian Ocean is not only infinitely smaller than it used to be, but the Indian mail, under the guidance of steam, has been granted almost a miraculous passage through the waters of the Red Sea.[4]

However, such euphoric receptions were met from the outset by skeptical voices and explicit rejection. It was not the people of the high-speed twentieth century who first complained that the status of the passenger was reduced to that of a package by the mechanization of the means of transport and the detachment of movement from the organic bonds (of the play of currents, winds, and waves; of carrying capacities; and of the fatigue and natural movement of animals and men).

Thus, the writer Joseph Conrad, who was a passionate sailing ship traveler and who was generous in his polemic against the new steamship culture, said in one of his last books that the only thing separating the travelers on the grand Atlantic ferries from the cargo stored below was the fact that the travelers walked a few miles a day on deck. A leading German media scholar, Bernhard Siegert, who was pursuing a research project on the ship as a site of modernity, added that what gives meaning to existence on board is "solely the maintenance of the itinerary."[5]

In his 1942 book *Land and Sea*, constitutional lawyer Carl Schmitt examined world-historical development as a territorial-historical panorama of "land and sea conquests." He explained how the hegemony of the British Empire was possible in the modern period: domination at sea. According to Schmitt's analysis, the prerequisite for this was an "exclusively maritime existence." The British Isles had to turn away from their own soil and become a ship, if not a fish: "The ship could hoist anchor and drop anchor in another part of the world. The great fish, the leviathan, could set itself in motion and seek out other oceans." The power of the British Empire lay in the sea and spread into a network structure across the globe. "The English world began to think in terms of bases and lines of communication. What to other nations was soil and homeland appeared to the English as mere hinterland."[6]

The medium of this spread was the sea, and the prerequisite for British power was control over shipping lines (and later also over electronic communications lines, the undersea intercontinental telegraph cable, which united all dominions of the United Kingdom into what was called the *all red system*). Thus, according to Schmitt, although England stood at the forefront of the development of industrialization, this development also marked the beginning of the end of British world power. Continental powers could attain a dominant position in that moment when "the leviathan was turning [from a giant fish] into a machine," because the principles of modern land transit were transferred to sea transit, and the categorical difference between land and sea disappeared.[7]

This difference had been the basis for England establishing its supremacy as a mobile island. Furthermore, aircraft, radio, and radar technologies, and later rockets and satellites—new means of transport and communication dependent on neither land nor water—gained central importance as strategic instruments of spatial mastery. In addition, England was no longer an island protected by a "moat," or at least not at an advantage compared to the continental powers because of its insularity. From the "ocean of sky," every landing site is an island, whether in the middle of a continent or on a ship.

An argument for a shift in power relations, as if it were necessary and driven by the laws of nature (and not the war machinery of nations), is certainly not unproblematic—especially one penned by an intellectual accomplice of Nazism in 1942. Schmitt's analysis concerns various forms of mobilization and their respective understanding of space. However, when reread in the context of the media and industrial-historical debates regarding the great changes of the time, such terminology gains new relevance.

Since the intensive discussion of globalization in the last few years, it has become clear that mobilization—the setting in motion of the means of production, capital, and people—is not only a warlike measure carried out by national armies but also a phenomenon characteristic of modern societies. The supporters of these great movements were from the beginning (even at the time of Schmitt's texts), and still are, largely nongovernmental, economic, and internationally, multinationally, or supranationally active agents.

At the end of the twentieth century, the American military historian and fleet captain Alfred Thayer Mahan noted in his great study of the history of sea power that the steamship had transformed the sea into a "system of highways" for the worldwide circulation of goods: "The sea, or water, is the great medium of circulation established by nature, just as money has been created by man for the exchange of products."[8] Therefore, according to Mahan's analysis, the purpose of military dominance was first and foremost the securing of economic interests.

Sea-Land's vision of a sea highway; advertisement from the *Journal of Commerce* 7, no. 4 (1966).

Consequently, the prime objective of a military strategy at sea must be to control the sea routes—and this is no different under modern conditions than it was at the time of the rise of the British Empire in the seventeenth century. However, with the introduction of mechanical sea travel, the essence of the sea changed. Without the dependence on the natural factors of wind, currents, and waves, the sea was transformed into a smooth surface. Seen from the perspective of military strategy, it had become a kind of ideal land without obstacles, reduced to a coordinate grid.

The decisive difference between the classical economy of the sailing ship, which made leaving land possible, and the modern economy of the steamship, which organizationally moved the sea or sea travel closer to land, lies in propulsion. Whereas a sailing ship could be on the water for a potentially unlimited time (as long as it managed to feed the crew) and thus become a kind of sea creature, a steamship is existentially dependent on its fuel supplies, without which it is adrift at sea.

As a result, the British Empire was compelled to establish a global network of coal stations. Secluded islands in the middle of the ocean gained central strategic importance as fuel depots (a structural development that was repeated with the development of the global airway network under the rule of the United States at the end of World War II and in the years that followed). This is how England temporarily achieved its most dominant position and showed the other nations—one last time—how to maintain a hegemonic system under modern conditions.

However, it was presumably precisely this shift to a land-based economy that bore the seed of Britain's fall from supremacy as ruler of the seas, because the decisive innovations of the end of the 19th century and the first half of the twentieth century all strengthened a continental perspective, a land-based perspective. Other nations therefore pushed their way to the foreground.

In the nineteenth century, the United States had already established a world system on the English model, but on a continental basis. This was possible because of the nation's enormous geographical expansion and substantial technical progress. Through the industrial development of inorganic chemistry, through fertilizer and synthetic fibers, broad self-sufficiency became possible also on a smaller level around 1900. The perspective of industrial production was able to shift from a geographical limit, dependent on the availability of certain raw materials, to a vertical limit on the increase in productivity. In this context, the American economic geographer Peter Hugill talks about a "German variant of the world system," since the decisive developmental steps in agriculture and the chemical industry were taken in Germany.[9]

In addition to modern chemistry, there were revolutions in energy production and distribution (electricity) and in motorized transport (cars, trucks, tractors, and aircraft) that made possible an increasingly intensive use of available space and a sustained increase in industrial productivity. As the network of maritime transport became still more dense, and transport volumes increased in the decades after World War II, in part because of containerization, this took place on a substantially different basis than the developments in the nineteenth century did. It was the expansion of *continental* economies that broke the categorical distinction between center and periphery, between coast and hinterland, brought about by the limitation to water-based means of transport, and changed it into a comprehensive arrangement of zones of various intensities of production and distribution.

Thus it is quite logical that the containerization pioneer and globalizer McLean would be anything but the kind of "old salt," heroized by Carl Schmitt, who opened up the "sea element" by leaving land behind.[10] McLean did precisely the opposite. He didn't reduce the land to coast and hinterlands. On the contrary, seen from land, the sea is little more than another somewhat softer stretch of highway. Without respect for its grand tradition and its historical importance, McLean regarded sea travel as a new field of business and nothing more. By shipping along the coast, he hoped to create a more economical alternative to the highly contested internal market between trucking companies and railways.

In mid-1955, the Pan Atlantic Steamship Company registered its new container liner service with the appropriate authority, the Interstate Commerce Commission, stating that it would employ ships as "sea tractors" in order to carry road transport containers. At least in retrospect, it is no wonder that Malcom McLean was not able to connect with traditional shippers with this idea and instead had to found his own company.

Initially it was only in the banker Walter Wriston that he found a partner. Wriston was head of the National City Bank, which financed the Waterman Steamship Corporation deal. Later Wriston would become the chairman of the renamed Citibank, which he led to become the largest bank in the world in the 1980s.

When Wriston recalled McLean, the banker said that the shipping entrepreneur was a naturally gifted financial talent. Apart from the highest standard of cost-efficiency, which served as the basis for McLean's decisions, there was a structural quality shared by the business environments of the two men: containers and currency. Both are metaoperators of circulation that smooth differences, create connections amid separation, and treat unequal things identically.

On the Liquefaction of Land

Simply writing about the drying up of the sea would not do justice to the complex developments of a globally connected world under the conditions of capitalism. At least to the same extent that the sea is subject to the organizational principles of land, the previously firm ground is made liquid, and territories are inundated with global currents: of goods, people, money, ideas, and belief systems.

By Friedrich Nietzsche's time (1844–1900), when steamships began to connect the "old" continents with the "new" several times a week, and when the undersea telegraph cables were being laid, complaints began to spread about the loss of firm (metaphysical) ground, at least in the secularized centers of Western society. Nietzsche turned the tables and encouraged the philosophers with an often-cited maxim to go "to the ships." However, he added a warning to this prompt. Whoever would set out into the unknown could never return—the loss of land would be final:

We have forsaken the land and gone to sea! We have destroyed the bridge behind us—more so, we have demolished the land behind us! Now, little ship, look out! Beside you is the ocean; it is true, it does not always roar, and at times it lies there like silk and gold and dreams of goodness. But there will be hours when you realize that it is infinite and that there is nothing more awesome than infinity. Oh, the poor bird that has felt free and now strikes against the walls of this cage! Woe, when homesickness for the land overcomes you, as if there had been more *freedom* there—and there is no more "land"![11]

Confined to the open sea in Nietzsche's post-Christian world, stable conditions have been plunged so far into crisis, have slipped so far from current consciousness, that they exist only in memory—and even there only in an unclear reference. His modern vision of the loss of foundation is a fairly precise negation of the vision of drying up portrayed in the biblical book the Revelation of John. Here the sea is dried with the destruction of the "whore of Babylon," to which all the irregular and uncontrollable currents of goods, people, money, and ideas flowed, the "waters" of "peoples, and multitudes, and nations, and tongues."[12] Instead, only the dry, clear conditions of the earth prevail under the single law of God.

However, for the atheist Nietzsche, the "land" is sunken and lost, the bedrock to which the primacy of a law, a binding complex of meaning, is necessarily tied; it exists only in memory (and that is a good thing). Furthermore, this memory is so vague that it can be referred to only in quotation marks, since everyone conceives of it differently and because, consequently, anyone can claim to have 'solid ground' to offer.

One of the protagonists in Herman Melville's last novel, *The Confidence Man: His Masquerade*, comes to feel this when he becomes the victim of an evil parody of the prophecy of John. The setting of the story is the Mississippi River, an area of the world where the indistinguishability of land and sea has continued unabated since the first day of creation, as the river carries along enormous volumes of mud and constantly changes course. On board a steamer, a confidence man, whose job consists of building unfounded trust, gives a student the chance to invest in shares of New Jerusalem, a new city that, although it is situated directly on the shores of the Mississippi, is supposed to be "terra firma" (firm ground). The only evidence the confidence man can provide is a written diagram.

In this story, New Jerusalem—the city where, according to John's promise, the endless mixing, groundlessness, and gluttony of Babel's economized culture will be at an end and only the law of the one God will reign—is itself based on fluid ground, on land that is regularly indistinguishable from water—if it is based on anything at all. Nothing more than lines on a page authenticate its earthly existence. "One cannot escape the suspicion that the coming kingdom of God is simply Babylon in disguise," Bernhard Siegert wrote in an analysis of this sequence.[13]

Incidentally, Terra Firma Capital Partners is the name of one of the world's largest private equity firms, managing gigantic funds and multiplying the assets with investments in various types of venture capital. Since the time of Nietzsche and Melville, the problem of a world of floating, nonreferential signs, the creation of value from nothing, has been further intensified.

Consider the increasingly abstract or derivative funds that constitute a large part of trade in the international capital markets (and the junk bonds, speculations on high-risk securities, which led significantly to the global financial crisis in the fall of 2008).

Consider the meaning of images distributed en masse, whose referential status has been indeterminable at least since digitization. Consider the fluidity of digital money, permanently processed in global data centers, and the total amount of which, if it could be calculated, would far exceed the equivalent value not only of actual goods but also of first-order abstractions, of money distributed as paper and coins.

Containerization's contribution to this development has been significant. On the one hand, it relativized the idea of the site of production to an extent not seen previously, since the boxes contain mostly intermediates, parts of parts, which are first shipped around the globe several times and handled in the most varied places before they come to market as fully assembled final products. On the other hand, container transport itself effected major abstraction and created signs. Since the boxes remain constitutively closed, nowhere in the entirety of the global transport process can you see what is inside. Rather, provided with the necessary knowledge and technical equipment, one can merely read the visible and invisible codes or enlist the help of elaborate imaging devices like scanners or X-rays.

Although Malcom McLean never really left land, instead trying to reduce the oceans to a system of highways—to dry up the seas, in essence—he also contributed considerably to the liquefaction of land. The transport of goods in containers largely levels the difference between sea and land transport. That is, the channels are opened in both directions. Whether one sees the seagoing container lines as extended land routes in a liquid element or surface routes as intracontinental channels between the world's oceans ultimately becomes a matter of perspective.

Bridges Out of Water

In reaction to the closure of the Suez Canal (1967–1975), and strengthened by the oil crises in 1973 and 1979–1980, container transport firms began to offer *land bridges* within the United States in the 1970s. These were container transports between either western Europe or the East Coast of the United States and East Asia that did not travel through the Panama Canal but were instead moved by train from coast to coast, saving the time and, above all, the fuel required to travel around the continent.

This concept got a decisive boost in the mid-1980s from the epochal shift in the focus of the world economy from the Atlantic to the Pacific region. Since then, the majority of the most significant harbors in the world have been on that side of the globe. In the United States, the ports of Los Angeles and Long Beach, California, superseded the ports of New York, for a long time the world's largest. On the other side of the Pacific Ocean, the harbors of Hong Kong, Singapore, Busan (South Korea), and Kaohsiung (Taiwan) gained rapidly in the 1980s. Shanghai, Shenzhen, and other Chinese harbors have now taken the lead.

Beginning in 1972, the firm Seatrain, which had already made its debut in 1929 with the intermodal sea-land transport of freight cars (the direct precursor to containerization), offered a service called *minibridge*, whereby containers were shipped from Asia to a port on the West Coast and, for a combined fee, transported by train to New York. Later, this offer was expanded to a complete land bridge—that is, a sea-land-sea transport chain from Asia to Europe and back. A mid-1970s magazine advertisement for Seatrain's intermodal offer stated the following:

At Seatrain, we move containerized cargo in a *functional* world, not a conventional geographical one. So why not show water routes through the U.S.A.? Our rail-and-sea international land bridge functionally allows a container to travel between Europe and Asia as if North America were cut by a 3,000-mile canal.... And why not show the continents closer than they used to be? Our new *Euro-class* containerships have clipped at least one full day off the time it takes to move goods between America and Europe.... We took a fresh look at today's needs, concluded that the old rules no longer applied. So we threw the rulebook out the porthole. In an industry steeped in tradition, our innovative ways of doing things may be controversial at times.... No wonder they all call us The Cargonauts.

Sea-Land

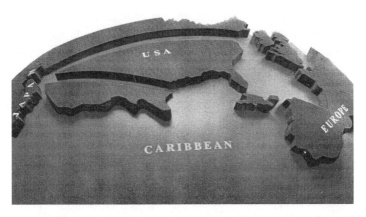

Transport shapes the world: continents crossed by container channels close ranks. Seatrain advertisement, late 1970s.

The Argonauts, a group of superheroes from Greek mythology, rowed their ship *Argo* through the Mediterranean, the Sea of Marmara, and the Black Sea to Colchis, in modern-day Georgia, and stole the Golden Fleece. According to one variant of the myth, they made their escape by way of streams, rivers, and inland waterways, carrying their vessel, if necessary, from one waterway to the next, through the continents of Europe and Asia and rowed their way around the land mass back into the Mediterranean. The late-twentieth century "Cargonauts" are not dependent on rivers and need not carry their ship. The containers used for the Cargonauts' means of transport changed from sea to land transport and back with relatively little effort. Nevertheless, their advance had other obstacles in its path, particularly legal ones.

In the United States, two deregulatory legal initiatives cleared the way for shipping transport companies to offer an integrated service of intermodal sea and land transport. The 1980 Staggers Act removed rail transport from the regulation of the Interstate Commerce Commission, which had governed it since 1935, and the 1984 Shipping Act opened the internal U.S. transport market to international shipping companies. In the wake of this liberalization, entirely new contract models were possible, and the prices for transport services fell considerably in the ensuing years.

Besides the customers of the transport companies, the greatest beneficiary was American President Lines (APL). The enterprise, already boasting a more than century-long tradition of international merchant shipping, established its own railway transport service in 1980, APL Linertrain. APL's president, Bruce Seaton, who played a similarly important role in the history of containerization in the 1970s and 1980s as Malcom McLean did in the 1950s and 1960s, explained in hindsight the move toward consistent organization of the entire multimodal transport chain from a single source. In a discussion with shipping historian Arthur Donovan, Seaton stated that it was found to be necessary to assume control over the entire transport process, on the high seas as well as on the continent.

Serving as background were the bad experiences that the APL had had in the winter of 1977, when severe snowstorms paralyzed railway lines across the United States and the railway companies tasked with container transport were unable to say where their trains or their containers were. In order to prevent such situations in the future, automated and computerized booking and monitoring procedures were introduced. Going along with the deregulation and further liquefaction of container transports in sea-land transit was the implementation of a control paradigm in which the increase in spatial freedom of movement and flexibility was paired directly with an intensification of the control of movement.

The most crucial hardware innovation for dealing with a task that a contemporary commentator referred to as "the greatest organizational challenge in the history of surface transportation"—namely, the "containerization of domestic America"—was the 1981 introduction of double-stack railcars, or *stacktrains*.[14] These are lowered railways cars on which two shipping containers can be transported, one stacked on top of the other. Malcom McLean had already made the suggestion in 1967 to several railways companies to utilize such cars, but with no success.

After McLean retired from his position as director of Sea-Land in 1977 and sold all his shares a year later, his successors revived the idea, and together with Southern Pacific Railway they developed a system for the transport of double-stacked 35- and 40-foot containers. APL followed suit in the same year. Since 1984, trains have been running that are put together entirely with these double-stack railcars. They can transport up to 300 40-foot containers (or 600 20-foot containers), and at this scale they can become competitive with coastal transport and continental circumnavigation by container ship.

The presence that these trains have in the United States today is demonstrated impressively in American filmmaker James Benning's 2007 work *RR*. The title is the abbreviation for *railroad* and is typically found on warning signs at crossings. The film consists of 43 static settings in various places, in most cases on flat land, in which a train traverses the frame in its full length and duration. Except for showing two passenger trains, the scenes

Agent of liquefaction: an APL stacktrain crosses the Mississippi River in East St. Louis, Illinois. Film still taken from James Benning's film *RR*, USA, 2007.

depict freight transport—from the archaic, classic, closed car to rail tankers, car transport, and double-stack container trains. Their share increases over the course of the film, and a historical development becomes visible.

In a departure from his previous films, Benning allowed additional material to be added over the audio in *RR*, and thus included commentary and semantic connections. In a key scene he established the link between the modern process of globalization and the biblical story of the apocalypse. In a panorama of a steep-sloped canyon where a spectacular railway line has been built, a train passes through the shot from right to left. Gregory Peck's voice is heard, reading from the Revelation of John about the "whore of Babylon" and the mixture of land and water done under her watch.

The boom of railway container transport in the United States led to a surprising reversal of circumstances, at least from a European perspective. The railway system, which had played a significant role in the development of the North American continent in the nineteenth century and had once been the largest and most modern in the world, stood more or less on the verge of bankruptcy after World War II and no longer seemed competitive in the face of burgeoning street transport.

However, the railways in the nations of continental Europe had always had a certain status as "state railways," despite similar structural problems, and were able to preserve a comparatively high level of social appreciation. Once the container system returned to the American continent as a transport system that took the path from the streets and rails to the seas and oceans and back, the railroads in the United States, reincarnated as "container roads," regained immense importance in the transport of goods.

In contrast, efforts to introduce a similarly efficient continental container transport system for intra-European transit and for transport between Europe and Asia have thus not come very far, which is why the old dream of an industrialized Silk Road is taking shape very slowly—if it will be realized at all. Too many regulations and too much resistance from individual nations have blocked agreement on the common policies and funding arrangements that would be necessary to operate pan-European railway corridors.

Though planned for decades in relevant bodies of European commissions, the corridors need double-stack container trains with hundreds of TEUs in transit in order for the financial investment to pay off. Nevertheless, motivation for new efforts could come from Asia. In the massive, booming nations of China and India, stacktrain lines have come into operation in recent years, and hundreds of miles of new track are under construction or in planning. (Since 2011 the "Yuxinou" train has been operating, transporting containers over a distance of 6,400 miles from Chongqing in mainland China to the German inland port of Duisburg up to three times a week. But its capacity of a maximum of 51 TEUs is still far from the economy of scale needed to make it an efficient land bridge ready to compete with ocean transport.)

With the introduction of a systematically intermodal service like the land bridge, the categories of land and sea blend together in both directions. This is reflected in the history of the companies involved. Perhaps a traditional shipping company like APL founds its own railway division. Or perhaps a consortium of venerable railway companies purchases a container shipper, as CSX did in 1986 with the Sea-Land Corporation (which was founded with the aim of being part of a group of sea and land transport companies but which then failed because of the business laws of the time).

The ownership of the materials used also expresses the mixing of categories: containers, railway cars, or tractor trailers could belong to a company with an emphasis either on land transport or on sea transport. Today the natural combination and mixture of sea and land transport in the concept of transport as a chain shows itself in the new structure and self-definition of formerly sector-based transport companies such as Deutsche Bahn AG (the former German Federal Railway).

As a logistics service, these companies naturally integrate all areas of transport under one roof. Thus, the range of maritime thinking identified by Schmitt for the British Empire, with the

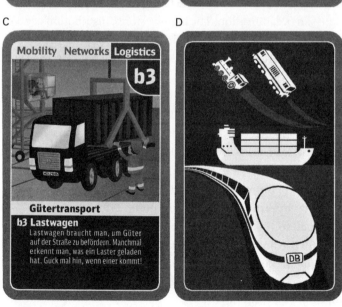

Shift from state railway to global logistics enterprise. Cards from "Oli's Train Quartet," Deutsche Bahn AG advertising gift, 2006. © Deutsche Bahn AG/Titus Ackermann.

simultaneous solidifying of the sea into land, extends to the continental land masses between the oceans. The organization into outposts and routes has spread across land and sea.

The intermodal container switches, now called *terminals* rather than *harbors*, stretch into the depth of continental space, where land transport systems form interfaces that differ only marginally from the land-sea interface. Storage and loading functions that were previously concentrated almost exclusively in port cities emerge along highways and around airports. Port cities fight against becoming pure "container channels," Hamburg regional economist and infrastructure expert Dieter Läpple said already in 2000.[15] Yet the rural and intercity areas converted into logistics zones must come to grips with a rampant dislocation—an oceanification. The big-box stores, distribution centers, large parking lots, storage facilities, and container depots distributed according to worldwide business plans look the same everywhere.

Seen from the sea, land can no longer be reduced to a strip of coast and hinterland (even if it is still common in maritime circles to incorrectly speak of "hinterland connections"). A view from the sea has become a special, functional perspective of the world, wet or dry (or in the air). Where this perspective prevails, liquefaction is produced. As philosopher Peter Sloterdijk writes, "all cities have meanwhile become ports . . . for where cities have not gone to the sea, the sea comes to them."[16] This is true not only for the flow of information. The "sea" flows both in lines of data and in transport. One must only be between the condensed urban zones to see that the waves of global commerce also break on shores far from the sea.

Around the World

If you take a regional train in the center of Germany, in Thuringia (where I resided when I wrote this book), from medieval Erfurt into classical Weimar, you will unexpectedly find yourself in a container situation. About halfway, near a small place called Vieselbach, multiple stacks of shipping containers stand along the railroad tracks. The boxes bear the logos of global transport. A gantry crane erected above the tracks and storage area stands ready for transfers. The cause of this unanticipated manifestation of a world system in the historical core of German regionalism is the IKEA Distribution Center–Central Germany—a large blue and yellow box building that the furniture corporation constructed several years ago, conveniently located along the railroad and the A4 highway from Dresden to Frankfurt am Main.

Would Malcom McLean, the architect of a world system yet still bound to his homeland, have reckoned on the resounding effect of the container idea? As a trained shipper, at least, he was aware of the continental depth that this combined land-water system would have to reach in order to work according to his vision. His love for North Carolina also did not prevent him from establishing terminals in all the eastern U.S. states. That was at the beginning of the 1950s.

By the end of the 1970s, after McLean resigned at Sea-Land, he purchased U.S. Lines from his stock proceeds. This was a freight shipping company with a tradition of passenger shipping that was threatened with failure amid the vast restructuring brought about by the introduction of the container. McLean planned the construction of a fleet of extremely large, relatively slow container ships that would circumnavigate the globe. (This idea is now, as a reaction to declining energy resources, being taken up again, such as with A. P. Møller Maersk's new Triple E Class ships.) McLean called these vessels *jumbo econships*. They were configured for 4,400 TEUs, roughly 50 percent more cargo capacity than the largest container ships of the time.

After the land bridge, so-called around-the-world services were the next conceptual answer to the oil crisis and increasing trade imbalance with Asia. Rather than regularly experience the problem of insufficient cargo volume common with shuttle transport back and forth along a route, these would instead travel in only one direction, such as from New York to the ports of western Europe, through the Suez Canal to Singapore and Hong Kong, to Japan, to the American West Coast, and back to the home port through the Panama Canal. From the United States eastward they would carry goods for Europe and for the growing demand of the Middle East (increasingly wealthy from oil dollars), and from the ports of East Asia they would carry goods for the American market. In addition to McLean's U.S. Lines, another company, the new Taiwanese firm Evergreen Marine, began a similar service.

McLean envisioned opening a global bus service, but his plan did not work. Global stagnation in the wake of oil prices, as well as overcapacity, put container shipping in a crisis at the time. The econships were too large and too slow to operate profitably. In addition, oil prices did not climb to the extent that McLean had anticipated, but rather they fell again in the course of the 1980s.

Evergreen, with its smaller and faster ships, succeeded and rose to become one of the world's largest container shippers on the basis of its more flexible global services; McLean's plan, in contrast, brought U.S. Lines to a spectacular collapse. In November 1986, the company was forced to file for bankruptcy, $1.2

billion in debt. Once again, McLean made history: at the time, it was the largest corporate bankruptcy in American economic history.

With this, McLean shamefully faded into the ether. Nevertheless, this failure did not cause long-term harm to the container transport business that he was so fundamental in creating, nor to his reputation as founding father of containerization. However, it remains open to question what importance the maiden voyage of the *Ideal X* in April 1956 had, given that this date is viewed today as the founding date of container shipping, and what role McLean played in economic and technological history, even though legend credits him as the "inventor" of the container.

Many historians have suggested that McLean should by no means be regarded as the inventor of container transport (let alone of the container), since there was already intermodal container transport over land and at sea decades before. However, this may be countered by the fact that none of these approaches consequently pursued or even conceived of the production of a combined sea-land transport. By transferring an element of land transport to sea shipping and producing the necessary technical interfaces for the optimally smooth change between land and sea transit, McLean, on the one hand, exploited the legal difference between sea and land enshrined in contemporary regulations: sea transport could demand lower tariffs than land transport, since it is slower. On the other hand, he also relativized this difference: his initiative led to the eventual fall of all tariff regulations.

This distinguished the sea-land system from earlier container transport concepts, which left the categorical difference between sea and land untouched and consequently continued to treat the harbor as the central point of meeting for maritime and continental transport. The credit certainly goes to McLean for being the first to see transport as a chain of land and water conveyance and for implementing this vision in the face of existing legal and economic conditions, as well as ushering in the construction of the relevant technical media.

For shipping historian Arthur Donovan, McLean was not the inventor of containerization, but he was certainly the inventor of the container ship, since the first full container ships were designed and built under his aegis in the 1960s.[17] In contrast, economic historian Marc Levinson emphasizes the importance of the conceptual idea. McLean was the first to see that it is the business of the transport industry to transport goods, not to operate ships (or trains or trucks). From this fundamental insight, a concept of goods transport was derived that differed from all previous transport concepts, because it made the container into the central component of a transport network.[18]

As for the evaluation of McLean's visionary power and "revolutionary" virtue, there may nevertheless be an inconsistency of some importance—indeed, one between the "container idea," the legendary core of McLean's own historicization of the situation, and the documented sequence of historical events. For in 1955, after he had purchased the Waterman Steamship Corporation and his plans for a coastal transport service had begun to take concrete form, these plans included utilizing RoRo ("Roll on, Roll off") ships—that is, ships to transport tractor trailers complete with the driver's cab and chassis. This would have merely represented the establishment of a more consistent road-sea transport network and not been a "container revolution." Only at the last minute did he decide to use the containers rather than the entire trucks.

Thus, if McLean was indeed the first to conceive of and implement the container transport concept and should therefore receive historical credit, then one must at least note that he developed these concepts successively and somewhat ad hoc, since he was already mired in the concrete implementation of less "revolutionary" plans. However, the legendary heart of the container fable—according to which the seed of the idea had already been sown in the 1930s but simply had to wait until the right moment—is actually a myth of historiography, a primal scene planted after the fact in order to make the story sound better.

In the classification of McLean's historical importance, perhaps it is necessary to move another aspect more strongly to the foreground: his talent for convincing others of his ideas and projects. The spectacular takeover of the Waterman Steamship Corporation bears eloquent witness to this; so does the fact that McLean easily found investors for his around-the-world plan with U.S. Lines who were prepared to offer $570 million simply for the construction of 12 ships for an untested concept.

Whether he was a visionary or simply had a good nose for business opportunities, McLean was a true confidence man. And the success of the container idea is in no small measure the success of a marketing strategy that announced its most important claim in the very name of the service: Sea-Land. With a similarly impressive tautology as the container, which always offers itself up as an argument for its own superiority, the name of the pioneering containerization firm expresses with compelling clarity the concept that would reorganize an entire industry.

Advertisement for an intermodal container moving service, United States. From *National Geographic* magazine, April 1911.

4 Container Histories

The container as such is ancient.
—Walter Meyercordt

An advertisement from the April 1911 edition of *National Geographic* magazine praises the intermodal transport service as follows: "LIFT-VANS can be provided for immediate loading in any city in the United States or in Europe. Their use insures a minimum of handling, security for small packages, and least possible risk of damage."

In the attached photo, next to the caption "hoisting Lift-Van on board steamship," a large steel transport container can be seen being loaded onto a ship with the aid of a crane. The entire side of the container is emblazoned with the name of the company and a description of the services offered:

BOWLING GREEN STORAGE & VAN CO.

18 Broadway, New York
Trans-Atlantic and Inland Removals

Since 1906, this American storage and moving company offered continental and transatlantic transport service in steel containers 18 feet long by 8 feet wide and 8 feet deep. Silvio Crespi, the chairman of the Bureau International des Containers (BIC), wrote in his 1934 container manifesto (see chapter 2), that "in any event, furniture trucks or liftvans have long been used for the transport of household goods across large distances."[1] Could there have been a container company before the invention of the container?

There were also similar services at this time on the other side of the Atlantic. In England the railways began to offer intermodal land-sea container transport after World War I. Before the war, the containers were still the classic, large, seaworthy boxes carried to port on flat railway cars and transported by ship to their overseas destinations. Now wheeled automobile cranes or port cranes transferred mobile boxes from freight cars between ship and rail. In 1933, the four largest rail lines in England already had around 4,000 of these large containers with a capacity of up to four tons, in addition to nearly 1,000 refrigerated boxes for meat transports. The large containers moved not only between railway and ship but also between railway and truck. Thus, this represented an intermodal transport system of rail, road, and water, though at a comparatively small scale.

Through a conceptual and technological change arising fundamentally from the conditions of railway transit, the traditional overseas box became a modern container, a transport medium *between* modes. Two central elements of container transport were already provided with these container services. The organization of a complete door-to-door transport was delivered by various means through a single supplier. And the amalgamation of an intermodal transit network of rail, road, and ship utilized a land transport container, the trailer, as transport medium and—at least conceptually—a system of intermodal *switches* arranged deep within the country.

For transports overseas: CONTRANS container. Door to door—low transport costs—cargo protection—small packaging effort—easy handling. Advertisement for the German Federal Railway, Norddeutsche Lloyd, and Hapag's Contrans container, early 1950s.

In 1952, Norddeutscher Lloyd, Hapag, and the German Federal Railway formed the Contrans Company for Overseas Container Transport to promote the overseas transport of goods in large containers. At the beginning of 1956—that is, before the premiere of the *Ideal X* (see chapter 3) in U.S. domestic transit—the two shipping companies offered containers of 176, 246, and 351 cubic feet for liner service to the U.S. East Coast.

The Birth of Container Transport

Why is it that none of these technological and organizational innovations from the first half of the twentieth century came to signal the birth of container transport? Why was the container "invented" in the United States after World War II, according to the more or less unanimous view of its modern historiographers? And how can one speak at all of a date of invention for a container, or with any other product with a complex technical and social development? Does a development not necessarily feed off of dozens of sources? It seems above all that the aesthetic necessities of historiography are to blame—the need for representation, for identifiable figures, and for narrative arcs—when complex, temporally, and spatially broad contexts of creation retract to a point and a date and find their sole beginning there.

With the Lift-Van system, there is clearly the idea of door-to-door transport with various vehicles and a single transport container, but from a modern perspective it lacks the standardized technical elements that so greatly accelerated the transfer and linked the intermodal transport process into a stand-alone system. It lacks the special container cranes and mounting devices; it lacks additional specialized container accessories, like corner fittings, twist locks, and spreaders; and it lacks stackability. The same may be said in principle for the English development between the wars and for Contrans.

Nevertheless, it may be said of the continental European development, as I described it in chapter 2, that it indeed led to the establishment of the first international interest group with the aim of standardizing container transport, and these standards also gained a certain degree of validity in international transport among various countries within Europe, but the actors were uninterested in ship transport. The European intermodal transport container was primarily a matter for railways, which were reacting to the challenge of the superior flexibility of trucking in land transit.

Regardless of the fact that sea transport was also mentioned in the statutes of the BIC, land and sea transport continued to be regarded as separate spheres, with the port as the border. Consequently, an entire series of technical solutions were proffered at that time (as well as during the nineteenth century) for the shift of cargo units from street to rail transport. But the change from land transport to ship retained the classic loading methods, with laborious mooring of the heavy containers and equally cumbersome stowing through hatches into the cargo hold.

Thus, from the perspective of the central elements of the modern container system, it seems plausible that the various approaches to container transport systems in the first half of the twentieth century did not yet represent the beginning of containerization. Nevertheless, such argumentation has one decisive blemish: it projects the present state as the result of a necessary development in the past and therefore undercuts all the coincidences, improbabilities, and complexities that shaped the concrete historical progression and that led to the formation of a technical system that appears as it does at this particular moment.

With a certain probability, the initial spark could also have led to a mechanized container transport system at another time in another place. Or an entirely different container concept could have become universal. Other intermodal concepts could have prevailed that did not rely on the container as a general medium but that had one vehicle accepting another, as do the classic RoRo ("Roll on, Roll off") ferries.

Certain problems of capacity and efficiency arose in many places in the wake of the industrialization of production and the globalization of the markets since the nineteenth century, just as there were certain technical and organizational solutions that seemed to impose themselves. However, for a relatively long time—at least measured by the speed with which the container system established itself worldwide after it found its current form—these various solutions remained latent, producing only local experiments or initiatives limited to certain routes, agents, or areas of application.

This observation could equally be made after the date now recognized as marking the "invention" of the container. After Malcom McLean's company Sea-Land brought the container to market as a universal intermodal transport container, it was a good 10 years before it actually proved to be a seriously competitive model for those providers of classic kinds of transport, and once it was recognized as such, it still faced fierce resistance. Thus, for a long time, the container remained a niche offering and might well have disappeared from view again, like other container systems before it.

Even in the early 1960s, European shipping companies were issuing enormous contracts for the modernization and replenishment of their transport ship fleets with conventional loading techniques. The 1965 annual report for Norddeutsche Lloyd pleaded for an "evolutionary development"—that is, a parallel expansion of cargo and container transport with combined ships in combined docks. This was despite the fact that the Bremen company, in cooperation with its archrival Hapag, would found the first European liner service to the United States with a full container ship only two years later. New York City also wanted to prepare for such a wholly anachronistic (from today's perspective) model of mixed transport when, in 1956, it decided on a gigantic expansion program for its port facilities around the Manhattan peninsula.

By the time of their completion in the early 1960s, all these facilities were already obsolete, as we will see. If New York City bosses, the traditional harbor freight companies, and the powerful dockworkers' union had succeeded in enforcing their interests, history would have taken a different path. It was no different in European ports. When Malcom McLean and Frans Swarttouw, his local agent, hosted a reception to mark the completion of the first container-processing facilities in Rotterdam harbor in 1966, they were booed by the Dutch shipping dignitaries. And in Hamburg, today by far the largest container port of Germany, a united front of politicians, warehouse companies, and port contractors was so resolute against the container that the first container terminal was not brought into operation until 1968, two years after the beginning of liner service between the U.S. East Coast and the ports of Rotterdam and Bremen.

The Three Phases of Container History

Thus, the question of when the container system got its start and when and why it expanded does not seem to be answerable with the naive clarity that talk of the container's invention in 1956 would seem to indicate. Consequently, I suggest a three-phase model for the history of the container and its prehistory, in which the period of the modern container's "invention" is a liminal period (as conceptualized by German historian Reinhart Koselleck for the indeterminate period between the last decades of the eighteenth and first decades of the nineteenth century as the beginning of modernity) stretching over several decades.

The first phase began with the first recorded transport vessels in the Neolithic period, with initial highlights (concerning quantity and spatial distribution) in antiquity and the Middle Ages, and then lasted into the nineteenth century. In this phase there was a relatively large variety of container forms, for the most part normalized, often even with a prototypical standardization, scaled to the capability of human (or animal) bodies.

The second phase was the liminal period just mentioned, from approximately the mid-nineteenth century to the end of the 1960s (thereby also encompassing the first 10 years after McLean brought his idea to market). This period witnessed the concept and the technical requirements for the modern container as the standardized medium for the mechanical processing of transport within an intermodal system—that is, one that integrated various carriers. However, these technical innovations found application only in relatively local contexts, limited to certain corridors.

Only in the third phase, from approximately 1970 till today, did container transport consistently expand as a sea-land network, to become the dominant form of (general) cargo shipping, with the relevant effects on and participation in the organization of production in the process of globalization. Strictly speaking, one would have to further divide this phase into two segments. From the end of the 1960s to the beginning of the 1980s, container transport by ship spread across the entire globe. The maritime side of world container transport came into its own. In the early 1980s, continental expansion began on a large scale, and with it came the closing of this system. Globally, the production of goods is aligned with the possibilities of the logistical system. The distinction between containers (spaces) for storage and those for transport is nearly erased. The result is the total mobilization of inventory. The broad distribution of production sites and the short-term, demand-driven manufacture of even complex goods (just-in-time production) became general principles.

Following the outline of the three phases of container history that I have proposed, I will begin at the start of recorded culture and illuminate from there some possible strands reaching into the twentieth century that may aid in the reconstruction of some prehistories of the container.

Culture from the Container

With their settlement in the New Stone Age, the time of the Neolithic revolution, humans began to produce containers such as vessels for the ashes of the dead (urns) and funerary goods, containers for supplies (jugs), and containers for transport (baskets). Humanity's first synthetic material, ceramic, was widespread by 6500 BCE. Consequently, the last segment of the Neolithic era before the transition to the Copper Age is called the Ceramic Neolithic Age (from sometime between 6500 and 4500 until 2000 BCE in Europe).

This age viewed the archaeological designation of entire cultures, eras, and regions according to the prevailing types of containers that they produced—or that were discovered in the greatest numbers during archaeological excavations. These include the Funnel-Beaker Culture (ca.4200–2800 BCE, central and northern Europe and southern Scandinavia), the Globular Amphora Culture (3100–2700 BCE, central and eastern Europe), the Bell-Beaker Culture (2600–1800 BCE, southern, western, and central Europe), and the Urnfield Culture (1300–800 BCE, central Europe). It would be very much in the spirit of this practice of naming and classification if our culture ultimately came to be called the Container Culture.

The American historian of technology Lewis Mumford developed an almost heroic notion of the importance of the container for the development of humanity in his great 1961 study of the history of the city, which drew a line from the first human settlements to the modern city. Neolithic society's change in focus from aggressive "masculine" cultural practices of hunting and warfare to the protective "feminine" ones of rearing and nurturing allowed the Neolithic period, an "Age of Containers," to become the cradle of our civilization, according to Mumford. He continued:

> It is an age of stone and pottery utensils, of vases, jars, vats, cisterns, bins, barns, granaries, houses, not least great collective containers, like irrigation ditches and villages…. Without tight containers, the neolithic villager could not store beer, wine, oil; without sealable stone or clay jars, he could not keep out rodents or insects; without bins, cisterns, barns, he could not make his food keep from season to season. Without the permanent dwelling house, the young, the ill, and the aged could not be securely kept together nor tenderly cherished. It was in permanent containers that neolithic invention outshone all earlier cultures: so well that we are still using many of their methods, materials, and forms. The modern city itself, for all its steel and glass, is still essentially an earth-bound Stone Age structure.[2]

Mumford extrapolated a whole series of sites and institutions from Neolithic container creations that were of central importance for the later rise of cities, some of which endure to this day: granaries, banks, armories, libraries, warehouses, irrigation ditches, canals, water reservoirs, moats, water supplies, and sewage systems, which "are also containers, for automatic transport or storage." Without them, Mumford added, "the ancient city could not have taken the form it finally did; for it was nothing less than a container of containers."[3]

The thesis of containers as creators of culture is supported by the myths of various peoples. There is hardly a cosmogony (creation story) that does not have a container metaphor. As a refuge for change, maturity, and reproduction, vessels are often equated with wombs. To mark his curated exhibition World of Vessels—from Antiquity to Picasso, housed in the Ludwig Galerie Schloss Oberhausen museum in 2004, art historian and container specialist Peter Pachnicke wrote the following:

> In the cooking vessel, raw things were transformed into cooked things, corn became bread in ovens, grapes became wine in barrels, seed became fruit in planting pots, human beings are formed in the womb through mysterious processes of blood-milk transformation. If you read the myths, legends and fairytales of the various cultures, you find that the fabled change processes for humans, for nature and the cosmos take place in containers drawn from the elementary experiences of the everyday. Cooking pots, jugs, cups, bowls, vases become vessels of fertility, birth, death, rebirth, wisdom, calamity, and wrath.[4]

Amphora and Barrel—Premodern Containers?

All the mythical container tales may be traced back to cultural practices of transportation or preservation—or more pointedly, from a media-technology perspective, as practices of transmission or storage—that have arisen since the Neolithic period and in the period of classical antiquity. In the course of their transmission history—from the Greeks to the Romans, from antiquity to the Renaissance (often by way of the Islamic cultural sphere), and from antiquity to modernity—errors in translation and continuity have crept into the texts, partly through the direct path of excavations and partly through the intervention of Renaissance texts and artworks. Eloquent evidence of this mechanism is the stubborn confusion of *pithos* (or *amphora*), the portable counterpart to the large stationary jug, with barrels or containers such as jars or urns in many English translations. For instance, in Homer's *The Iliad* as translated by Richmond Lattimore, there is a discussion of Zeus's vessels (*pithoi*), which are always bulging and which assign fortune and misfortune:

There are two urns that stand on the door-sill of Zeus. They are unlike for the gifts they bestow: an urn of evils, an urn of blessings. If Zeus who delights in thunder mingles these and bestows them on man, he shifts, and moves now in evil, again in good fortune. But when Zeus bestows from the urn of sorrows, he makes a failure of man, and the evil hunger drives him over the shining earth, and he wanders respected neither of gods nor mortals.[5]

This vessel myth is very interesting, because it provides the archaic motif of the cornucopia with significance by dividing it into good and bad contents; it thus establishes a tradition of fatal vessels that is critically important in the modern era, to this day. The large containers "on the door-sill of Zeus" can actually only be *pithoi*. These were distributed throughout the entirety of Greek antiquity—bulbous clay storage vessels with considerable capacity. They were used for the storage of grain, but they originally served as urns as well—as vessels of the underworld in the burial of the dead. A similar situation exists with Pandora's box, which also must have been a *pithos* in the Greek myth. And the Cynic philosopher Diogenes's "barrel" could not have been such, since the technology for barrel production was simply unknown during his lifetime.

It seems obvious to find the basis for such confusion in the apparent or actual historical comparability of amphora, barrel, jar, urn, and container as transport vessels of different ages.

Thus, the (Roman) amphora, as the first "modern" transport container, seems to be differentiated from the eponymous vessels of earlier container cultures by the fact that its appearance as a means of transport for an empire is *not* limited to a locality or region. (Nevertheless, this argument is weakened by the fact that it may also be reversed. Because we recognize the Roman Empire as the prototype for modern empires and as a form of globalization, we subsume entirely regional differences in the container forms under one category). What is undisputed is that remains of amphorae may now be found in all areas that were part of the Roman Empire, or that were at least in its realm of influence. They are the most widespread and the most often discovered artifacts of antique Roman culture.

Antique and modern transport vessels share another element: coded inscriptions. The discovery of shards of Roman amphorae is of such great historical interest not least because the ancient transport containers, in contrast to other vessel discoveries, generally have a written code. From the impressions stamped into the containers before firing or the characters drawn on the completed products (*tituli picti*), inferences can be made about place of production, ownership, contents, and transport routes. These are comparable to modern forms, in which the type, origin, and quality of the contents as well as its path of transport could have been controllable.

But is it really justified to speak of amphorae as the most important transport medium and symbol of antique world commerce, or to call the barrel the container of the Middle Ages? Just how far can we go with the comparability of the amphora, barrel, jar, or urn and the modern container? To answer these questions, it is necessary to look a bit more closely at the forms of ancient and medieval trade and goods transport. Even if the differences

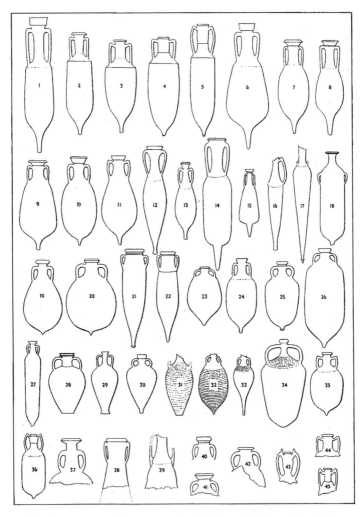

Table of Roman amphora categories in Dressel's *Corpus Inscriptionum Latinorum*, 15, Pars 1, Berlin, 1899.

Typical markings on Roman amphorae. Nos. 1–8 from *Roman Amphorae* by M. H. Callender, published by Oxford University Press, © University of Durham 1965. From D. P. S. Peacock and D. F. Williams, *Amphorae and the Roman Economy: An Introductory Guide* (London: Longman, 1986), 10.

are greater than the commonalities, the comparison of premodern and modern means of transport should at least bring out the specific characteristics of each with respect to the other.

The History of the Amphora

Vessels from Canaan, documented since the fifteenth century BCE and used from this period in trade with Egypt (as shown in Egyptian reliefs), may be seen as direct precursors to the amphora. The amphora apparently arose from the need for a type of container that was large enough to transport bulk goods but could still be carried by a person. With the two handles somewhat above the center of the vessel, the weight could be well-balanced. The pointed base made it stackable and therefore particularly suited to transport by ship in larger quantities.

In the course of the eighth and seventh centuries BCE, Phoenician merchants spread the amphora into the western Mediterranean. In the seventh century BCE, it gained a foothold in Greece and, through Greek trade, in Sicily and southern Italy. There, at the turn of the third century BCE, the Greek-Italian amphora type came to be, from which the Roman amphorae would later evolve.

At the time of the Roman Empire, Mediterranean trade had reached a significant volume, even by modern standards. Thus, since the first century CE, the capital city of Rome was supplied with food largely by imports from the provinces. Wheat and other bulk solids were transported in sacks. Of wheat alone, 80,000 to 150,000 tons were shipped annually from Egypt to Italy, corresponding to 240 shiploads. Amphorae served in the transport and storage of the two other most important goods of the day: olive oil and wine. The same was the case for fish sauces, preserved southern fruits (dates and figs), marinated olives, and honey. Contemporary cargo ships could take up to 10,000 amphorae on board, with an estimated payload of up to 450 tons. The main purchasers of these food supplies were the inhabitants of the city of Rome and the members of the Roman army.

Shards of a certain type of amphora from southern Spain—named Dressel 20 by the German archaeologist who undertook the first comprehensive inventory of Roman amphora types—are found throughout the territory of the Roman Empire, particularly in its western provinces. These amphorae held nearly 20 gallons of olive oil and weighed nearly 70 pounds by themselves. They were stable and fit for sea travel, and compared to other smaller types of amphorae, they had a better ratio between dead weight and carrying capacity. The shards are dated from the first century BCE to the third century CE.

During this time, there was a tremendous boom in olive cultivation and olive oil production in Baetica, modern-day Andalusia. The oil was used not only for cooking but also for grooming, for medical treatment, and as fuel for lamps. It was brought to the Baetis (now Guadalquivir) River in skins and then bottled into amphorae at a central filling station under the control of a representative of the Roman Empire. These were marked with a stamp, a kind of early bill of lading, that noted the origin, the weight of the amphora, the weight of the contents, the date of inspection, the name of the inspector or the inspection number, the name of the owner, and the name of the merchant.

Then the amphorae were shipped downriver to the sea. Each amphora was generally only used once, and when it was emptied, it was thrown away. A massive mound of waste in Rome known as Monte Testaccio, consisting of more than 80 percent Baetica amphora shards, attests to the scope of foreign trade in amphorae.

From Amphora to Barrel

Despite the vast appearance of shards, one must also recognize that the use of amphorae was limited not only to certain goods but also to certain means of transport. For instance, skins were generally used in the era of the Roman Empire for the transport of liquids over land. This explains the professional history of the *utriclarii*. These ancient "movers," whose name is derived from the Latin term *uter* ("animal skin sleeve" or "leather sack") and who have existed since the first century CE, earned their keep by the transport of wine and oil from the producers to the next

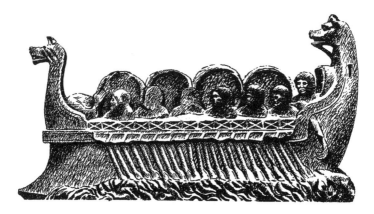

Ship with wine barrels on the Rhine. Relief from the Neumagen region, third century BCE. From Dieter Hägermann and Helmuth Schneider, eds., *Propyläen Technik-geschichte Band 1: Landbau und Handwerk 750 v. Chr.–1000 n. Chr* (Berlin: Propyläen, 1997), 256.

city or bottling point as well as between cities and regions. Wall paintings in Pompeii show how wine was transferred on-site, to the wine merchant or restaurant, from the *utriclarius*'s wineskin.

As wine production moved from southern Gaul northward over the course of the second and third centuries CE, winemakers and dealers increasingly utilized barrels for storage and transport. The *utriclarii* reacted to this by also loading their wagons with barrels. They transported wine to the south and brought back other goods on their return trip. Moreover, they increasingly entered trading themselves. Thus, regional wine and oil transporters became merchants over great distances, and the barrel became their most important transport medium.

The barrel established itself as the main means of transport through the northern and western provinces during the first centuries CE. In the process, the Romans adopted a Celtic-Germanic cultural practice. Compared to the amphora, the barrel was a large container. It had the additional advantage of being able to be rolled and was less fragile. Nevertheless, barrels could be used only on a limited basis in the more southern regions, since sun exposure dried out the staves and produced leaks.

In the following centuries and to some extent to this day, not only fluids were stored and transported in barrels but also a variety of produce in need of protection from desiccation and spoilage, as well as goods that required protection from humidity: butter, herring, salt, sugar, fruits, and pickled meat. Because of its versatility and general use, the barrel, as we noted earlier, has been called the container of the Middle Ages. The barrel was of central importance for medieval trade. The history of modern seafaring would also be unthinkable without supply barrels.

Barrels are assembled from premade parts: wooden staves and the iron hoops that hold them together. This connects them not only with the modern container, which is fabricated with profiled and sheet metal, but also with the ancient amphora. The research into their production is preliminary, but one can say, at least for the larger types of amphorae, that they were produced in multiple steps. As the base was being spun on the potter's wheel, several premade spiral rings were gradually affixed to it and partially fired. Only at the end were handles applied and the final rim shaped.

Ancient amphorae were also already produced serially and in large quantities. Here, however, the parallels end, because a core structural element of the modern transport container is its standardization. Supranational bodies with representatives from business and state administrations meet and negotiate binding guidelines for dimensions, materials, permissible technical solutions, and so forth.

Nothing of the sort can be said for amphorae or barrels. The barrel certainly served as a kind of large universal container in the Middle Ages and in modernity, as least for a diversity of produce. It was also stackable and reusable, and in this respect it exhibits structural similarities to the container. However, barrels were made for centuries with the same technique, albeit in very distinct regional sizes.

A comparatively large variety was also evident among ancient amphorae. In contrast to the scale of their modern iteration, the scale of the ancient transport and storage containers was geared toward what humans and animals were capable of moving. Certain standards or types developed locally and established themselves in regionally defined areas. Apart from the Dressel 20 amphora type mentioned earlier, dozens more were identified.

Presumably, the fact that these found such enormously wide distribution has to do with, more than anything, the great popularity of the product in them: the olive oil from Baetica. Thus, since the amphorae were bound to a certain product and normally used only once, despite the size difference it may perhaps be more logical to compare them with something other than the container—another genuinely modern transport and storage vessel: the tin can.

Toward the (Modern) "Container"

"Most often, vessels don't succeed, just the resulting container."[6] This is how art theorist and philosopher Hannes Böhringer paraphrased the difficulty of arriving at holistic and enduring concepts under the conditions of modernity in the age of the container. Since the rise of the modern, industrialized container principle, a criticism of objectifying and dehumanizing conceptions of space has come about that is based on the notion of the (soulless) container (and related ideas), while it conversely brings into play the vessel ("Gefäss") as the epitome of premodern wholeness.

Interesting, however, is that this usage represents a smooth etymological reversal. For as a glance in the German *Duden* dictionary will show, the term *Gefäss* ("vessel"), presumably derived from the Old Germanic words *Vaz* ("barrel") and *vazzen* ("to hold"), carried the basic meaning well into modernity of "receptacle," as a generic term for portable containers in the household, farm, or church. In contrast, the term *Behälter* ("container"), comprising everything from a barrel to a bag in today's parlance, pointed etymologically to a specific context in its common meaning: large, immobile places of containing, such as fishponds, houses, or castles.

A keyword history—research into the changing meanings and fields of meaning of the words *Behälter* and *Gefäss* in lexica from the eighteenth to the end of the twentieth century—clearly shows the change that the terms brought about, corresponding to the formation of modern social and knowledge systems taking place at the time, of logistical thought and of a new, largely nostalgic interest in the objects of the premodern past.

Volume 3 of Johann Heinrich Zedler's the *Great Complete Universal Encyclopedia of all Sciences and Arts*, in print since 1732, did indeed announce under the keyword *Behaltung* the modern general meaning of a container for the purposes of storage and transport, but the modern sense of mobility was derived from immobility: "dwelling, receptacle, stable, barn, house, farm chest, trunk." Only the last two meanings, the "farm chest" (presumably a large chest that was often put in the hallway in farmhouses) and the "trunk" (a "travel box, generally with a rounded lid and leather covered with fur"), referred explicitly to mobile containers. However, under the keyword *Gefäss*, one finds the general meaning of an object to be filled (with fluid): "Vessel, vase, ... 1.) refers to any instrument into which one can put liquor, juices and other things."

The 1854 edition of the German dictionary *Deutsches Wörterbuch*, by the Brothers Grimm, confirms this reading of *Gefäss*. Here, meanings from the realm of practical usage are preeminent, like loading, carrying, decoration, equipment, and attire. The word *Gefäsz* was already being used in the Middle High German period as a general term for drinking vessels, bowls, plates, and so forth: "the term also extends to containers as a whole, which is true to its origin; if one does not know or want to use the particular name, the term *Gefäsz* will do, now as in the sixteenth century."

In addition, the passages cited above refer emphatically to the fact that the term is closely tied etymologically to the business of loading and shipping—in other words, to transport—although the word nevertheless always refers to transportable, portable things. In contrast, according to the Brothers Grimm, the term *Behalter* refers to a divine attribute—*Jesus unser behalter (heiland)*, or "Jesus our keeper (savior)"—whereas the entries for *Behaltnis, Behältnis* give the impression of spaces rather than objects: "repository for goods, coal, clothing, animals, fish; repository for people, prison."

Nevertheless, throughout the rest of the nineteenth century, *Gefäss, Behälter,* and related terms not only lost their religious significance but also largely disappeared from the lexicon. *Gefässe* appeared in the first encyclopedia editions of the German dictionaries *Brockhaus* and *Meyers Konversationslexikon* as parts of plant, animal, and human anatomy. The term reappeared at the beginning of the twentieth century as dishes or cooking tools.

Yet in the sixth edition of *Meyers Grosses Konversationslexikon* (published in 1905), *Gefässe* were "prehistoric" vessels—that is, artifacts of past cultures.

In contrast, the term *Behälter* first reappeared when the logistical reformatting of industrial culture was already in full swing. In the seventh edition of the 12-volume *Meyers Lexikon* (published in 1924), there was originally no entry for *Behälter*.

But the supplemental volume 13 (1931) contained the following precise definition under the heading *Behälterverkehr* ("Container Transport"—in which the connection to the events following the World Automobile Congress is unmistakable): "*Behälterverkehr* (*Containerverkehr*): the transport of goods, first introduced on English railways, in trailer-like, closed, removable boxes that are moved onto ships or overland transport, for instance a truck chassis, so that the goods are moved onward, being reloaded more quickly."

This is the rebirth of the *Behälter* as container. The medieval holder for things in the human environment became the technical framework for goods, a global transport network of standardized containers whose primary goal consists of being emptied as soon as possible.

Paths to Intermodality

One of the core ideas that led to container transport is intermodality: the connection of multiple modes of transport—by land, by water, by air—into a single transport process. In principle there are three technical possibilities for realizing this linkage. The first consists of loading one means of transport entirely into or onto another, such as trucks on trains, railway cars on street transporters, or trucks and trains on ships. This so-called piggyback approach (which also includes the concept of RoRo ships) played a meaningful role before and during the development of modern container traffic and still maintains an important place in the transport business.

The second possibility lies in the construction of special modes of transport for use in several contexts: vehicles that can travel both by road and by rail, or some that can drive and float. As far back as the early history of the railway in the 1810s and 1820s, German rail pioneer Ritter von Baader envisioned a rail-street hybrid scheme under the title "System of Progressive Mechanics," in which road and track would function as complementary transport infrastructures that are traversed by the same vehicles.

However, like most of the other ideas in this direction, his design did not catch on. Indeed, the hybrid vehicle option is in line with a modern ideal that is still widespread, based on universal versatility and multifunctionality, but it has not brought about any results suitable for the system besides offering niche solutions, primarily for the military. Too complicated, too fragile, and too little payload—such was the judgment of Johann Culemeyer, another pioneer of intermodal transport, who developed a platform car 100 years later, inspired by Baader, which could transport railway cars or other extremely heavy loads on the street.

The third possibility for producing intermodality involved exchanging only one part of the vehicle (without the chassis, motor, or steering) between the different modes of transport—specifically, the part that holds the cargo: the container.

The engineers Reinhold Bräuer und Max Krusemark detailed how the decision for or against the various intermodality concepts was far from uncontested in a sketch that they included in their 1933 book *Jumping the Tracks: The Economic Battle between the Freeway and Railway—Container Transport*. It has to do with a system of relay vehicles in the intermodal transport chain. Similar to Baader, Bräuer and Krusemark position their work against heavy technology and exclusive infrastructural development in favor of the integration of all means of transport currently in use.

According to them, the basic mistake in the development of general transport containers was made in conceiving them as large, heavy boxes with flat bottoms for transport on a flat railway car. Bräuer and Krusemark were instead convinced that the maximum scope of intermodal container transport, true door-to-door transport, could be achieved only by a rolling container, a "light, nimble vehicle." Consequently, they developed their model for the general transport container after the example of the tractor trailer. They stated the principle of their relay vehicle as "freeing the container from the notion of the box and giving it at least the beginnings of a vehicle."[7]

From today's perspective, it seems that the most remarkable thing about their recommendation—again similar to Baader's approach—is that they did not justify the need for increased transport efficiency through the bottleneck of increased freight volume; rather, they did precisely the opposite, justifying the need through falling average transport volumes, which calculations solely in truckloads made unprofitable. However, this is the answer to a question that is still being posed today, above all because of the growing diversification and specialization of production.

In contrast, the system of large containers was able to develop its superior profitability only under conditions in which heightened investment costs were balanced by a massive volume and continually increasing transport levels. In fact, the weight and capacity of the shipping containers as general global loading

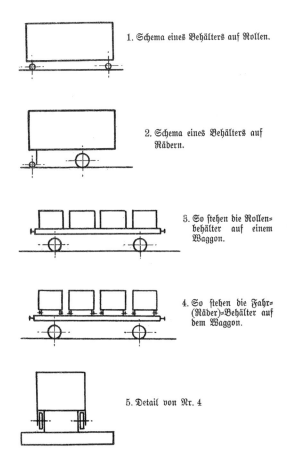

Sketch of a relay vehicle, Max Krusemark und Reinhold Bräuer, 1933. "The new vehicle has become a relay vehicle, which can move its cargo to a new vehicle in minutes without reloading, and this vehicle, which has the same handling characteristics as the first, then brings the cargo to its destination or to a third relay, etc." From Reinhold Bräuer and Max Krusemark, *Der Sprung aus dem Gleise: Der wirtschaftliche Kampf zwischen Auto und Reichsbahn—der Behälterverkehr* [Jumping the tracks: the economic battle between the freeway and railway—container transport] (Munich: Schweiszer, 1933), 38–39.

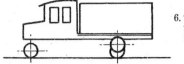

6. Das Stafettenauto mit aufgeladenem Behälter.

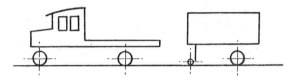

7. Das Stafetten=Auto mit abgesetztem Behälter.

8. Wer kein Stafetten=Auto hat oder den Be=hälterspediteur nicht nehmen will, holt sich seinen Behälter so: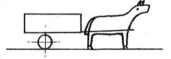

9. Oder so: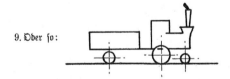

1. Diagram of a container on reels.
2. Diagram of a container on wheels.
3. This is how the reel containers stay put on a rail car.
4. This is how the wheel containers stay put on a rail car.
5. Detail of No. 4.
6. Relay vehicle with loaded container.
7. Relay vehicle with discharged container.
8. The ones who don't have a relay vehicle or don't want to assign with a shipper get their container like this:
9. Or like that:

units exceeded that of even the largest (railway) transport container in Bräuer und Krusemark's time four- to eightfold. Their introduction required that the world cargo fleet be converted to wholly new types of ships and that the ports be outfitted with a new infrastructure of the heaviest machinery tailored exclusively to the needs of the transport container. Ultimately, it was not the "light, nimble vehicle" that prevailed, conforming to the diversity of transport conditions, but rather the heaviest box and its dimensions that the production facilities and transport infrastructures had to incorporate.

The History of the Box

Like the historical shift in the meaning of *Behälter* outlined previously, this development, the decision against the vehicle and for the box, indicates another obvious way to bring the container from the depths of history: as a history of the box. On the one hand, it is a mobile carriage, a body separated from the vehicle's frame, and on the other hand, it is a box brought to the dimensions of railcars and streetcars.

Spurred on by the World Automobile Congress (1928), the founding of the Research Association for Container Transport (1928), and the International Container Competition (1931), a series of publications appeared in quick succession in the 1920s and early 1930s that dealt with both the historical predecessors and the existing systems of intermodal transport from around the world. From this history came the development of new ideas. The spark was the recognition that the railways, at least in the transport of package freight, and generally in the transport of smaller payloads, could compete with trucks long-term only if they catered more flexibly to the needs of their customers and, for instance, offered concepts for door-to-door transport.

Viennese railway engineer Fritz Brauner, one of the leading figures in the thought collective of railroaders working on concepts of intermodal and container transport, developed something like a general theory of the box in his attempt to historically and conceptually trace modern container transport. In a 1933 book titled simply *Container Transport*, first published in a series by the Government Committee for Economic Efficiency, Brauner suggested a classification system for transporters that would be applicable to all historical types of transport. It consisted of four broad type categories: load class, transport class, protection class, and size.

Load class included sacks, jars, support frames (platforms) and boxes. *Transport class* included lifting and carrying containers, rolling containers, driving containers without engines, driving containers with engines, transport units in water or air transport, and container parts for the handling of goods (such as jar handles or, today, twist locks on a container). *Protection class* consisted of container forms that serve to protect the goods. Here, however, the author did not name any specific container forms but merely recounted "the various possibilities for loss" that a container should prevent, such as leakage, theft, damage resulting from pressure, and explosion (to name a few). Under the fourth category, *size*, there is a table ranging from the "1st tier up to 2 kg retail packaging" all the way to the "23rd tier with 1,500 to 15,000 ton ship loads."[8]

Brauner classified the platform as the most important historical load class in land transport, since the cart and the wagon were developed from this, and thus the transition from carrying to driving. The box also emerged from the platform. Brauner defined the box as a container that can be "loaded under pressure (for instance when stacking the same container on top)." He continued, "The general box actually consists of six platforms, its most well-known representative is the usual shipping box."[9]

However, this historical and systematic classification of the box requires a certain restriction, because boxes or chests were also carved from tree trunks. This method of production is obviously older than that of the platform, since it can be done with a comparatively crude device. The etymology also speaks to this. The Middle High German word *Truhe* ("chest") presumably derives from the Old High German *truha* ("wood" or "tree") and *druha* (a hollowed-out stick or stone).

Antique vase imagery attests to the fact that carpenters in the time of the ancient Greeks already assembled chests from boards. Thus, we may conclude—at least for modern history but presumably also for antiquity—that chests, bins, and boxes were produced according to the platform principle. If one understands a platform to be a support frame with a flat surface affixed to it, as Brauner does, then this ancient principle also holds for the modern container. This generally consists of a base and a ceiling frame connected by four corner posts.

Because of their unaltered construction principle, on the one hand, and their functional provisions, on the other, the box—or, as the case may be, the box, the bin, and the chest—may be seen as direct historical precursors to the modern transport container. As Viennese folklorist Konrad Köstlin claims, the chest is dually coded within medieval and modern culture: one preserves one's most valuable possessions within it while also keeping them mobile.

The history of the cultural practice of storage and transport in the container–furniture chests spans from the medieval tradition of the dowry chest to the nineteenth century emigrant's chest, finding its continuation in modern door-to-door transport. Because chests were meant for such transport from the beginning, farmers' daughters came to the city with a crate of their belongings when they hired themselves out as maids. Move-in day at the new master's residence was called "trunk day" in northern Germany.

Male servants and students also traveled with their chests. The dowry that a daughter's family had to give was determined since the late Middle Ages by municipal regulations, and chests served as a measure. Should a marriage be ended because the husband died or because a divorce was granted, then the wife always had a right to her dowry chest and its contents.

"No object of mobile property is so abundantly represented in cultural-historical museums as the chest," Köstlin wrote.[10] Over the centuries, the chest has been the unit of storage across all social ranks and regions of Europe. Similar functions are also known for the chest from other areas of the world, such as China, and in some cases these functions are only now being lost in the wake of the country's rapid modernization.

In the European realm, the chest began to yield its leading role to the wardrobe and dresser in the seventeenth century, and with it its central utilitarian and representative function. The dowry chest survived only into the beginning of the twentieth century in the countryside. The development of the chest branched out. On the one hand, it led to the closet, itself a kind of house as a rooted and enlarged chest, which stood metonymically for the stability of the house and the circumstances of its inhabitants. On the other hand, it led to the trunk, traveling furniture that is used only when the inhabitants of the house with a closet leave their sturdy home for a time.

Furnishing Containers, Not Dowry Chests

What distinguishes the historical predecessors from the modern container? Chests and boxes have always served in storage as well as in transport. However, the container parts for the handling of goods described by the container theorist Brauner were geared toward human hands and not systemic technologies. There were grips, tabs, buttons, or handles, but no corner fittings or twistlocks, which ideally connect the transport containers with the processing network between the carriers without human intervention.

Furthermore, the box-shaped predecessors of the container were not general but rather were produced for a specific purpose and for specific kinds of contents. It is only in the course of the nineteenth century that precisely this connection was lost in some places. New, radically temporalized types of containers began to proliferate. Thus, the modern moving box and the Lift-Van container are mobilized storage and transport spaces that are utilized only in the exceptional state of relocation and not in the normal state of living.

Since the end of the nineteenth century, European railways have offered true door-to-door transport with containers. French and English railroad companies employed wooden transport containers to move household furniture. The boxes were reloaded between the flat railcar and the horse-drawn cart by way of cranes and then delivered to the customers' houses. A few decades later, during the Nazi era, the capacity of transport containers served to limit property. Jewish emigrants were permitted to take only one large container with them through the Lift-Van transport system, filled with their earthly possessions. The majority of these goods never reached their destination.

Even today, household goods in need of transport are still measured in terms of container capacities. For instance, it is typical at some American universities, as well as at many large businesses, embassies, international organizations, and so forth, to pay for an employee's relocation for professional reasons. The scope of the move is measured according to the capacity of 20-foot containers.

Boxes that were uniquely individual became common containers. Everything had its place in the box, in the beginning and forever, even while transitory, but precisely this kind of fixed order was lost with the implementation of the modern moving cartons. Alone, the exponential increase in the number of household items to be moved with each relocation today makes a stable mobile container arrangement impossible. The chest restricted the amount of one's possessions to what could be moved within it. The closet, the dresser, and the shelf, the chest's immobilized historical successors, opened up room for the exponential growth of inventory—and of goods to be transported in the case of a move, since it is not only the mass but also the number of stored objects that has increased. In other words, containers must also be transported separately from their contents.

The box is an "instrument of relocation," media scholar Claus Pias wrote in a wonderful miniature about the moving box:

[It] has its medial conditions in the form of normalized sizes, capacities and labeling areas. In a certain sense, it is the network protocol and invisible discursive condition of the move. All boxes look the same, are stackable and get mixed up with such reliability in the transporter only because it is their nature to some extent to form ahistorical spatial configurations. In the moving box, it seems that the transitory has come to itself and has come to the end of history.[11]

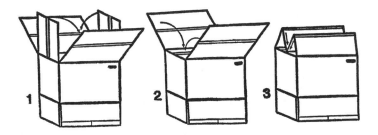

The most personal belongings in the standardized box format were the folding instructions for a moving container. From Hans-Rudolf Lutz, *Die Hieroglyphen von heute: Grafik auf Verpackungen für den Transport* [The hieroglyphs of today: graphics on transport packaging] (Zurich: Hans-Rudolf Lutz Verlag, 1990), 394.

In the chest, the history of the objects that it contained (and therefore the history of its owner) had been saved. In the box, all histories are reconfigured with each load. What lands inside it depends on chance, the organizational and logistical competency of its packer, and, ultimately, the standardized volume and capacity of the box itself. Stackable and optimally filling the transport volume of its vehicles, it ideally receives its measurements from the superordinate box, from the platform or the container. Standardization and arrangement into a logistical network of transport optimization allowed boxes to become containers: Containers in containers.

Aerial view of a Wal-Mart distribution center in Virginia. ©Wal-Mart.

5 Logistics: The Power of a Third Party

Diode, Triode, Method.
—Michel Serres

From the outside, the configuration looks like two enormous, flat boxes arranged diagonally. Almost hermetical, as if the entire structure consisted of two massive stone blocks, on closer inspection it becomes apparent that rectangular holes are cut into the walls at regular intervals at floor level. All around, hundreds of trucks with containers are parked. Roughly one-third of them are positioned with their back ends to the openings. One could get the impression that the trucks are box-shaped children being nursed by a box-shaped mother. And, in fact, the configuration functions a bit like this.

What was installed here, like a lunar base in the middle of the forest in Virginia, is a distribution center, one of many regional hubs for the Wal-Mart superstore chain. The "nursing" is carried out in both directions. One part of the arriving container trucks receives cargo, and the other delivers cargo. Along with the stationary infrastructure, the building, they form a highly efficient logistical arrangement.

Measured by its earnings and the number of its employees, Wal-Mart is not only the largest retail chain but also one of the largest businesses in the world (ranking third, down from first, in Fortune 500's list of the world's largest corporations in 2012). If its earnings were compared to gross domestic product (GDP), it would be the 27th most powerful nation on Earth, close behind Argentina and just ahead of Austria (according to the United Nations' list of countries by GDP for 2011).

On average, a container destined for Wal-Mart arrives in an American port every 45 seconds. Historically speaking, the company can be compared in its riches and influence only with the world-shaping enterprises before the rise of nation-states, like the German Hanseatic League in the fading Middle Ages or the British East India Company and the Dutch United East India Company in the 17th and eighteenth centuries. Wal-Mart's global ascent began in the early 1990s. It coincided, on the one hand, with a liberalization of world markets after the fall of the communist systems, an opening up not seen since the nineteenth century. On the other hand, it was inseparable from the differentiation and expansion of the worldwide container transport system on land and sea.

The discount superstore chain, founded on flat Arkansas land in 1962, owes its success to an optimization of logistical chains from the production of goods to their distribution and sale. In his book *The World Is Flat*, journalist Thomas Friedman wrote with fascination about the coordination of streams of goods upon his visit to a Wal-Mart distribution center: "Call it "the Wal-Mart Symphony" in multiple movements—with no finale. It just plays over and over 24/7/365: delivery, sorting, packing, distribution, buying, manufacturing, reordering, delivery, sorting, packing."[1]

This new organizational thinking has been perceived by many to be a logistical revolution. Wal-Mart had a significant part in this, which is why the new economic relations brought about by this change are also called the "Wal-Mart Effect." The power in the framework of producers, transporters, and vendors of goods has shifted in the direction of the latter.

Before the logistical revolution, the producers determined supply by their production cycles in what was called a *push economy*, to which the retailers had to adapt through warehousing. The risk of whether their inventory would correspond to demand lay largely with them. In the *pull economy*, made possible by refined logistics and intensified transport, large retailers like Wal-Mart give orders to producers and transporters on the basis of customer data and precise knowledge of their holdings. In so doing, they both reduce their warehousing costs to a minimum and determine what goods are produced when.

From Factory to Distribution Center

Today, all large retail chains operate with a business model similar to that of Wal-Mart. The company's importance is demonstrated spatially from the proliferation of its box architectures. On the one hand, there are so-called *box stores*, usually erected in the logistics zones of large cities: the spaces linking the suburbs, highway interchanges, and business parks. They are large windowless buildings, and most are painted with the company's corporate-designed colors and emblazoned with a giant logo so that they can be recognized from a distance. On the other hand, there is the ever more densely woven network of logistics centers: the distribution centers, or *distriparks*. They function as "mediators between the global system of harbors and ships

Automatic package processing in a Wal-Mart distribution center.
©Wal-Mart.

and the regional systems of trains and trucks," as architect Fred Scharmen wrote in an analysis of their function.[2]

A regional logistics center should have sufficient storage and, above all, processing capacity to continually provide all the branches within its catchment area with goods on schedule and in keeping with changing demand. Its interior is organized like a gigantic computer whose processing units are boxes. The task that this computer building was constructed to solve has to do with a problem typical of modernity: minimizing the number of paths traveled and simultaneously maximizing the amount of goods transported or, in this case, processed.

The mathematical foundation of this principle was already laid in 1736 by Leonhard Euler with the Königsberg Bridge Problem. The famous mathematician, who had just accepted a position as professor in St. Petersburg, wanted to know if it would be possible to cross each of the seven bridges in the city of Königsberg (modern-day Kaliningrad) that led over the Pregel River exactly one time and still return to the starting point. It is not possible. But Euler's solution, which entirely disregards the actual distance and road curvature and instead relies exclusively on the schematized connecting lines between the bridges, serves as the foundation of graph theory. In its topological (and no longer topographical) treatment of paths, only corners, edges, connections, and crossings are decisive.

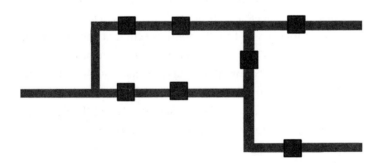

Schematic depiction of the Königsberg Bridge Problem: Is it possible to cross each bridge exactly once along a single path? © Alexander Klose.

Such a model's abstraction of physical space and the underlying mathematical formulation of spatial and temporal problems of distribution form the basis of modern logistical thinking or computing. The Traveling Salesman Problem, first mentioned in 1932, is also constructed on this basis. Here the sequence for visiting multiple places must be chosen so that the total distance is the shortest. The same is true for the Chinese Postman Problem, examined in 1962 by Chinese mathematician Mei-Ko Kwan, which attempts to determine the shortest path for a mail carrier who delivers letters on both sides of a street. Both are combinatorial optimization processes that may be applied in a multitude of fields, from microchip production to tour planning to genome sequencing. They are at play both for the analog-driven cargo distribution of a classic shipping company from 40 years ago as well as for the computerized process flow optimization of a modern, fully automatic container terminal.

The distribution center is the point of the highest technical refinement of the logistical system. With frantic speed and with a mostly automated process, containers are unloaded, goods are packed and labeled, and pallet loads are assembled. At the same time, logistics in the distribution center at the beginning of the 21st century has come back around to one of its starting points in the early twentieth century: the factory. Only now the goods are not manufactured—that is, produced from beginning to end—but are merely collected, assembled, or lightly modified.

In the distribution center, there is an apparent change from Fordism, the cost-cutting mass production of one product, to a flexible specialization characteristic of today's system of consumer capitalism. Products become a combination of modular components, sets of basic types with minimal variation, from which the buyer must choose. A product is often not even

Logistics

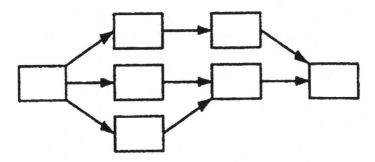

"Network, vectored, valuated and contiguous graph without loops for the modeling of project flow." From Jürgen Bloech and Gösta B. Ihde, eds., *Vahlens Grosses Logistiklexikon* [Logistics dictionary] (Munich: Vahlen, 1997), 738. Used with permission.

completed by the company whose brand name is emblazoned on the item. Instead, adjustment to individual customers' needs, seasonal demand, and labeling and packaging are outsourced to logistics services. The distribution centers function as central distribution stations and manufacturing sites for an increasing share of the objects of the modern consumer world.

Rationalization

Thanks to the efficiency of the container system, transport costs have sunk so low that in today's economic framework it is generally less expensive to have a product produced in various locations around the world, rather than at regional manufacturers, and have it transported tens of thousands of miles. From complex industrial products like an automobile to comparatively simple household items or toys, goods emerge from a network of producers that spans the globe. The big retailers, though citing the need of customers for lower prices, largely determine for themselves when and under what conditions a product is made.

According to the modern understanding, *logistics* essentially means that a logic of cost reduction and of flexible product design, grounded in the (supposed) needs of the customers, is effective in all links of the economic chain. It is no longer labor and production that are at the heart of things, but rather purchasing power and distribution; the classical economy is turned on its head. How did it come to this reversal, in which a long-neglected area of economic action, transport, unnoticeably crept into focus and shaped the whole world?

The seed of this development lay in industrialization itself. It was according to the principles of modularization, standardization, mechanization, and automation that first production and later the distribution of goods were fundamentally transformed. All these principles had their antecedents in the nineteenth century, both in Europe and in the United States. However, it was the American pioneers of *scientific management*, Frederick W. Taylor and Frank B. Gilbreath, who formulated a program from these principles in the first two decades of the 20th century. According to Taylor, the program should be

> applicable to all kinds of human activities, from our simplest individual acts to the work of our great corporations, which call for the most elaborate cooperation, ... to the management of our homes; the management of our farms; the management of the business of our tradesmen, large and small; of our churches, our philanthropic institutions, our universities, and our governmental departments.[3]

It was automotive industrialist Henry Ford who implemented the new principles in his factories—and simultaneously popularized them in his writings—so convincingly that they became the model for the entirety of further development in

Container transport from the railway to the factory building of the Ford Motor Company, ca.1918. From Horace Lucien Arnold and Fay Leone Faurote, *Ford Methods and the Ford Shops* (New York: Engineering Magazine Company, 1919), 26.

all industrialized and industrializing nations of the world. The scientific disciplines of management and organization theory forming in the years between the two world wars—and substantially in Germany—made "Taylorism" and "Fordism" the core of their project of *rationalization*.

Precisely what is meant by this term, which to this day is part of basic managerial vocabulary (particularly in reference to staff reductions and the introduction of new technical systems of all kinds), is not clear, even after looking at numerous relevant sources. The formulation found in the *Concise Dictionary of Economics* (1981) is typical: "Replacement of traditional approaches by more expedient and well-conceived ones for the improvement of existing conditions." But how it is determined what is "more expedient," more "well-conceived" or an "improvement of existing conditions"? A worker or employee's view would presumably be different from that of a board member, who feels primarily obligated to shareholder value. According to an intuitive understanding, *rationalization* would mean reforming a process according to rational—that is, reasonable—criteria. However, what is meant by *reason* and *rational action* has already changed many times in the course of the history of the terms.

In the 1920s, "rationalization," "new objectivity," and "modernization" according to an American model were fads that were enthusiastically adopted in Germany. A utopia of rationalization prevailed, encompassing industry, architecture, state bureaucracy, art, and literature in equal measure. Admittedly, this was simply a "dream of reason," in historian Detlev Peukert's words.[4] There was a great deal of enthusiastic talk at this time in Germany, but little action, little technological innovation, and little structural change.

The implementation of the visionary programs first took place on a noteworthy scale during World War II, but particularly after it. As part of the "economic miracle" [*Wirtschaftstwunder*] of postwar Germany, it was fueled on an ideological level by the battle between two power blocs disguised as economic systems, both of which sought their salvation in rationalization. This was as true for the introduction of a standardized container transport system as for home construction, the automation of production, or the design of functional rooms like kitchens and bathrooms.

The use of the term *rationalization* lends a scientific veneer to the managerial discourse and the call for technical and organizational change. However, it is a rather empty word whose core is interchangeable and whose truths are self-fulfilling. It is comparable to those "concepts of movement" characterized by Reinhart Koselleck, the great analyst of "Basic Concepts in History" (as the English title of one of his most influential works reads), in regard to the "isms" of the eighteenth, nineteenth, and early twentieth centuries: patriotism, liberalism, socialism, communism, nationalism, Zionism, and fascism.[5]

Common to all these is the fact that they are only marginally (if at all) based on experience, and instead open up horizons for a future yet to be created, from which it can then be determined retrospectively what the original intent was. Rationalization, the modern belief in reason put into action, was a continuation of the grand historico-philosophical projects of the nineteenth century. Only seemingly free of ideology and with no less of a totalitarian claim, it provided a direction and a model for the course of all social processes.

Logistics is its methodological child. One can by no means say that everything brought about by logistics and rationalization is all bad. It's my interest here not to judge the achievements of the logistical project but rather to explore its intellectual-historical roots and its philosophical core.

The second half of the twentieth century lived out the dreams of the interwar period. Transport research, struggling for a planning-coordinating reorganization of the increasingly complex motorized transport business, which had already taken up the term *rationalization* since the 1920s, was also betting on the concept's promise of salvation. A 1962 article in the journal *Rational Transport* entitled "Rationalization" states, "The object of our efforts lies in promoting rationalization in the area of transportation."[6] However, first and foremost this means modularizing, standardizing, mechanizing, and automating the promotion and processing of goods according to the principles of Ford's factory.

Pictorial statistic from an American military logistics journal, 1950s. The graphic demonstrates the rationalization effects—that is, reductions in work hours—achieved by the use of pallets throughout the entire logistical transport chain, from production to the field. *Containers and Packaging*, Autumn 1952.

OPERATION	PALLETIZED	LOOSE CARGO
UNLOADING CAR AT SHIPSIDE	9 MAN-HOURS	24 MAN-HOURS
LOADING AND STOWING ON SHIP	37 MAN-HOURS	47 MAN-HOURS
UNLOADING SHIP TO DOCK	24 MAN-HOURS	164 MAN-HOURS
LOADING TRUCK AT DOCK	8 MAN-HOURS	54 MAN-HOURS
UNLOADING TRUCK AT SUPPLY DUMP	9 MAN-HOURS	57 MAN-HOURS
STOWING AT SUPPLY DUMP	9 MAN-HOURS	39 MAN-HOURS
RELOADING AT SUPPLY DUMP	8 MAN-HOURS	55 MAN-HOURS
UNLOADING AT POINT OF USE	9 MAN-HOURS	56 MAN-HOURS

TOTAL NUMBER OF MAN-HOURS REQUIRED FOR MATERIALS HANDLING AT ALL OPERATIONS :—

PALLETIZED CARGO .. 203 MAN-HOURS

LOOSE CARGO ..... 682 MAN-HOURS

NET SAVING EFFECTED BY PALLETIZED UNIT LOADS ... 682 − 203 = **479 MAN-HOURS**

A Plus Value ... PALLETIZATION GREATLY REDUCES PILFERAGE POTENTIAL AND DAMAGE TO INDIVIDUAL PACKAGES

"Logistics, or the Practical Art of Moving Armies"

Modern knowledge of logistics is based on the parallel structure, integration, and synchronization of streams of production, distribution, and information. Precisely because it merges formerly separate spheres and dissolves borders with the logic of fluidity, the notion of logistics seems to be of such great relevance that it has become extremely prominent in recent years, and not simply in the narrow realm of economic publications. The tremendous expansion that logistics has experienced and continues to experience as a branch of economics is reflected in a kind of reporting and contemplation that classifies retroactively under that term all that was formerly called transport, trade, or commerce. However, this ignores that the term itself has undergone a revolutionary and rapid historical development.

The word *logistics* comes from the Greek verb *logizomai* ("calculation, computation, reasoning, consideration"). This is based on *logos* ("speech, calculation, reason"), which is related to *legein* ("speaking, thinking"). Our word *logic* is also grounded in *logos*. As early as ancient Greece, an action of calculation that (in contrast to arithmetic) was directed not toward the nature of numbers but rather toward the solution of practical problems was called logistics. Even if there is no direct link between that notion of logistics and our modern one, it is nevertheless remarkable because it was a concept directed toward the practical implementation of mathematics that could certainly be seen as a precursor to the modern models of process control—a mathematics interested not in truth or beauty but rather in forming and optimizing processes.

In the business environments that it so heavily dominates today, logistics was first discussed only in the 1960s or 1970s, and even then only sporadically. Initially this referred to the more narrow area of transport, handling, separation, picking, and storage. In the 1980s, systemic and integrated concepts moved alongside this understanding, stemming from management science, organization theory, and operations research. Thus, these concepts have their origin in the first half of the twentieth century, but they are only now associated with the notion of logistics. The situation is similar to the dominant concept of *flow optimization*, which is pervasive at all levels of a business. Even though thinking in terms of circuits and streams, employed in the eighteenth century (and not only in the field of economics), belongs to the founding moments of modernity, it established itself in this specific form as a new logistical paradigm only in the last 20 years.

Military supply may be seen as a direct historical predecessor of modern transport logistics. Supposedly, the Byzantine emperor Leo VI (866–912) was the first to explicitly use the term in this context. However, in the most commonly used meaning today, the word *logistics* is derived from the French verb *loger*, suggesting the accommodation and provisioning of soldiers. In 1638, the French army created the position of *maréchal général des logis de la cavalerie*. This is the officer responsible for the supply and shelter of the troops.

The beginning of a genuinely modern military-technological concept of logistics lies with General Antoine-Henry Jomini. Through his writings, the organization of supplies for fighting forces moved to the front row of military technical knowledge. In his 1837 work *Précis de l'art de la guerre* [*The Art of War*], published in Paris, the Swiss military theorist described extensively how logistics was the third most fundamental element of military operations—after tactics (the art of decision in combat) and strategy (the art of waging wars on the map). The pertinent chapter is titled "Logistics, or the Practical Art of Moving Armies." Resuming and reformulating Jomini's instructions, Israeli military historian Martin van Creveld comes to the following general definition of logistics: "the practical art of moving armies and keeping them supplied."[7]

Although Jomini's reflections became largely forgotten in the German- and French-speaking realms because of other military-strategic prioritizations, the science of military logistics was further developed in the United States. An English-language translation appeared in 1854. This is how it came to be that the concept returned to Europe a century later as "logistics."

The modern principle of logistics, first theoretically defined in Jomini's commentary, essentially means the dissolution of a transport paradigm that revolves primarily around the overcoming of *space* in favor of a paradigm in which the control and coordination of *timing* is at the forefront. In the nineteenth century, the acceleration and stabilization of the spatial processes of transport through mechanization, the famous "killing of space" by the railways, created the conditions for this development. With the rise of a regime of timetables, the problem of the coordination of numerous transport processes or circulations came to the fore. Consequently, the media of transport became time-critical—that is, the question of whether the media function, in the context of the system—was decided substantially by whether they adhere to their allotted times.

To use a definition from Hamburg regional economist Dieter Läpple, logistics brings about a focus on chronological sequences, the "orientation of all functional areas of an enterprise or of a production network toward the 'primacy of the flow of goods and news.'"[8] In light of the application of this priority

Title page of the government journal *Containers and Packaging*, 1957. In the U.S. Army, the fundamental importance of logistics for modern warfare was recognized very early. Innovative logistical concepts played a considerable role in their wartime victories. Nevertheless, it was enterprises of the free market that advanced the technology and organization of the container system and the logistical revolution.

in the most varied functional areas, even beyond genuinely economic contexts and in view of the expansion of a business calculus with logistics at its core to all areas of life, it seems justified today to speak of a basic logistical order.

The Postal Service: Pioneer of Intermodality

Today, Deutsche Post DHL, the successor of the good old German federal post office, is the world's largest provider of logistics services. After its privatization, and following the acquisition of other companies like Danzas (formerly and currently DHL) in 1999 and Exel in 2005, the enterprise, previously organized like a government agency, transformed into a globally active corporation with more than half a million staff. Courier and express services like DHL, UPS, TNT, or FedEx—and similar providers of transport services for small shipment units, such as mail-order books or medication delivery—are firms that organize a transport chain from desk to desk.

Thus, even after and despite containerization, their main object of business has always remained the individual parcel. In addition to the large retail chains like Wal-Mart, they constitute the second aspect of the logistical revolution. With the formation of ever more sophisticated chains of transport with cargo well below container size, like boxes and envelopes, they serve as a paradigm. All this has a certain historical consistency, because the first agents of intermodal transport were courier services. And the first institution to expand this system was the postal service.

In principle, it makes no difference to each letter (or each papyrus) which path it follows to reach its addressee. The main thing is that it arrives, and on time. Thus, as part of the famous

Logistics

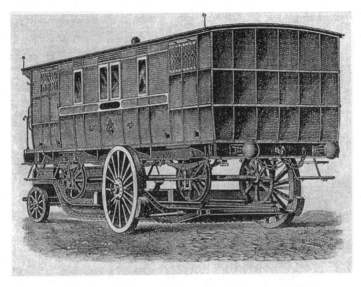

Vehicle for the transport of loaded railway mail coaches on the street to remote post offices off the railway line, Russia ca. 1870. From Johann Culemeyer, *Die Eisenbahn ins Haus: Die Beförderung von Eisenbahnwagen und Schwerlasten mit Strassenfahrzeugen* [The train to the house: transport of rail cars and heavy loads by road vehicles], ed. Alfred Gottwaldt (Düsseldorf, Germany: VDI-Verlag, 1987), 3. Used with permission from Springer-Verlag GmbH.

ancient Roman information system *cursus publicus*, speed ships were used to transport urgent letters from Ostia to Sicily, Carthage, Britain, or Spain. In port, they surrendered their shipments to mounted messengers.

However, there is a decided difference between the messenger and courier services that existed in all parts of the world when groups of people ruled over larger territories and the mail of the modern era, from which today's postal system emerged. The former were tasked with transmitting a specific shipment from a sender (for instance, a ruler or a high representative of a bureaucracy). The latter collect shipments from a plethora of senders and *organizes* their transport. For these, it makes no essential difference whether they carry out the transport themselves or have them transported by others.

Acting in this way, as an agent between transport carriers and purchasers of transport services, the contemporary post assumes a mediating position that will become typical of modern capitalism. In today's container system, they are comparable to *nonvessel operators* (NVOs), which solely undertake the organization of transport—as carriers without their own vehicles, shippers without their own ships, dock handlers without their own cranes and personnel. However, they are primarily responsible for services from the beginning to the end of the transport chain because they negotiate contracts with the various carriers.

The path to the modern postal service is characterized by general availability, regularity of service, networklike consolidation, intensification of postal connections in space and time, optimization of the processes of collection and distribution, and production of a generally valid tariff system. The actual breakthrough to the principle of intermodality, always inherent in the idea of the post, was preceded by several technical and administrative innovations. This happened in the first two-thirds of the nineteenth century: the introduction of railways and steamships that coalesced into the land-water network, in the transport of people as well as parcels; the introduction of consistent postage in the British Empire after the postal reforms of Rowland Hills in 1840, which served as a model for the reformation of postal tariffs in other nations; and the worldwide standardization of formats and tariffs in the wake of the 1874 founding of the Universal Postal Union.

As media scholar Bernhard Siegert has written, it is only now that "all means of transportation, no matter what kind—whether dromedaries, khaki-clad Ascaris, Trans-Siberian or Central Pacific Railways, carrier pigeons or steamships" are being integrated into a transport chain despite their individual qualities, for which the customer pays precisely one internationally standardized price.[9] The introduction of the postcard has contributed fundamentally to the establishment of the concept of a world postal service and a standard postage since 1869, because it not only required the necessary agreement on standards but also simultaneously brought its own format into circulation. As a kind of "virus of the world postal system [...] it forced old postal districts to adopt the norms of the world post and thus to succumb themselves,"[10] the postcard was, in a certain sense, the container of the nineteenth century.

In light of the general proliferation of electronic communication via the Internet, the end of the postal service is often predicted. Yet its astonishing renaissance as an avant-gardist of logistics corresponds in a certain sense to a return of packet shipping. Likewise, in the history of the business, this belongs in the prehistory of container transport. The precursors to modern container ships were packet boats. And the precursors to modern container shipping companies were packet shipping companies like the two most important German carriers, the Hamburg America Line (H.A.P.A.G.) and Norddeutscher Lloyd; England's peninsular and oriental steam navigation company (P&O), the largest and longest-standing imperial mail line in the United Kingdom of its day; the Dutch Netherlands Steamship Company and Royal Rotterdam Lloyd; and the French Messageries Maritime and General Transatlantic Company.

Since the mid-nineteenth century, these carriers operated liner service between Europe and overseas. They owed their ascent above all to colonialism, because in order to maintain their overseas possessions, the colonial powers were dependent on reliable transport routes for goods and information. Especially valuable or particularly urgent cargo was transported on the high seas in "packets"—mostly sturdy, oiled leather or burlap sacks. The transport of the post was a lucrative task, carrying a high amount of responsibility and assigned only to reliable, well-known companies.

Form Chains!

The idea of the container—the standardized transport unit *between* the carriers, neither vehicle nor cargo—is a product of the first, purely industrial phase of rationalization. At the same time it went beyond its scope. Thinking in terms of the container required thinking of hitherto separate spheres as one context, a *chain* reaching from production to the place of consumption. However, making this chain a reality involved an administrative, legal, and conceptual reorganization at least as much as it did technical or mechanical aspects, a fundamental revision of the relationship between internal and external contexts of transport or production. The systemic context that previously consisted only of the production of transport expanded in a relatively short time to the *organization* of the complete process of production.

Thinking in terms of the chain means reevaluating the *in-between* connecting the formerly separate spheres, bringing the power of a third party into play that we have generally grown accustomed to calling *logistics* since the last third of the

Logistics

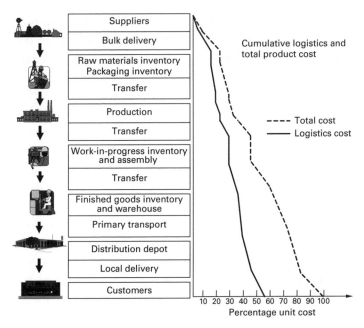

"A typical physical flow of material from suppliers through to customers, showing stationary functions and movement functions." Figure 1.4 from: Alan Rushton, Phil Croucher, and Peter Baker, *The Handbook of Logistics and Distribution Management*, 5th ed. (London: Kogan Page, 2014), 14. Used with permission.

twentieth century. Thus, the transport rationalization of the early years becomes globalization: the distribution of production sites across the globe according to the most favorable conditions, the mobilization of warehousing, and the proliferation and alignment of local markets. In other words, it becomes a reorganization of all areas of production and consumption in the transport chain. And its links are containers.

Container chains cut clear across oceans and continents. If one goes so far as to claim logistics to be a new business discipline, which today constitutes the core of economic thought according to popular perception, does it also follow that its development was necessarily linked to the containerization of transport? Is logistics necessary in order to investigate the scope of knowledge opened up by the time-space system of the container and to spell out and operationalize this organizational knowledge?

Scientific texts on transport from the early period of containerization certainly reinforce this thought. The founding protocol of the Bureau International des Containers (BIC) in 1933 already held that an intermodal transport container must be able to be "integrated into factories and warehouses as well," so that it could be "carried from the site of production to the user."[11] An article entitled "Transport Chains, Integrated Transit and Containers—Key Points in Business Logistics" that appeared in *Rationeller Transport* in 1969, at a time when the results of the implementation of the ISO container in intercontinental goods transport began to be visible, sums up the resulting need for "thinking in systems" in this way:

> Technological development in the area of goods transport and handling has now created a kind of "Deus ex machina" that nearly forces thinking in systems, in the container.... A containerized transport relationship can only reach maximum efficiency if ... the technical, tariff and organizational aspects of a relationship are brought to a common denominator. This insight led to the so-called "integrated intermodal transport chains," to transport chains that were based on a holistic concept, beyond diverse means of transport and terminals.[12]

Since the mid-1990s or so, the term *supply chain management* has shifted to the center of discussion. According to this,

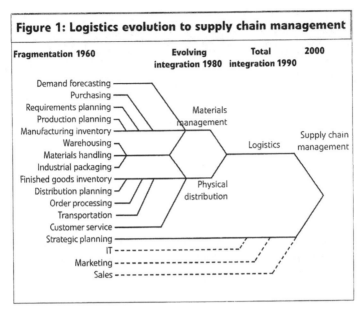

Graphical depiction of the development of logistics from 1960 to the present. It begins with a great "fragmentation." More and more areas of business are classified under the jurisdiction of logistics. This, in turn, is merged into the holistic idea of *supply chain management*. From Chris Thorby, "Freight Forwarders and Logistics: Evolution and Revolution," *Containerisation International*, 40th anniversary edition (2007): 21.

logistics includes the production and distribution activities of all businesses taking part in an economic process, from production to sale, including streams of information and money.

The fact that this concept is often illustrated by the supply chain of the food industry points to the "cold chain" as a civilian predecessor of the idea, and the term. In 1908, French engineer

Graphical depiction of the cold chain from the slaughterhouse to the household refrigerator in a German prospectus from 1935. From Ulrich Hellmann, *Künstliche Kälte: Die Geschichte der Kühlung im Haushalt* [Artificial cold: the history of domestic cooling] (Giessen, Germany: Anabas-Verlag, 1990), 148.

Albert Barrier suggested the term *chaîne de froid,* soon translated into English, for the organization of transport of refrigerated and deep-frozen food from the producer to the end user's refrigerator. The pioneer of this kind of food chain was the intercontinental transport of meat and fruit from Central and South America, taking place since the 1880s (thanks to the founding of the traditional German shipping company Hamburg Süd, among others) and increasing in scope and importance in the twentieth century.

Parasite

A chain is composed of separate links. Thus, somewhat paradoxically, thinking about a logistical chain must proceed from this constitutive separation of interrelated things. This is how the term *broken transit* occurred. Conceptualizing the interaction of various kinds of transport as a systemic network, leads to the realization that this linkage is permanently disturbed at critical points—namely, at its interruptions—where the different means of transport clash. Compared to actual transport, the handling of goods often costs many times more in terms of time and money. Consequently, the perspective is changed from functioning to nonfunctioning to disturbance, and one looks for a way to eliminate this disturbance, a bonding agent that makes the change in cargo more fluid.

The container is this agent. In the multimodal system, it works like a magnet, concatenating the metallic transports on land, at sea, and sometimes in the sky. This is how Klaus Ebeling, director of the European Intermodal Association, an organization for the promotion of multimodal transit based in Brussels, put it in a conversation I had with him in early 2005.

In essence, intermodality is the thought and organization of the in-between. This in-between is filled—physically—by a mobile bit of (transport) space bound in by a standardized hull, a black box that can be filled with and emptied of every conceivable content. Moving outward from this point, all participating transport systems and situations must reorganize, regardless of their evolved cultures: road transport, rail transport, sea transport, harbors, and terminals.

In this respect, the container can be regarded as a parasite, in the sense of French philosopher of science Michel Serres's media theory. It uses existing transport systems, acts on them, and compels them to transform. It doesn't meet them at eye level. Since it can't move itself, it climbs on their backs. *Piggyback* is the term for the intermediate model, the precursor to the container in which entire trucks, complete with their trailers, were transported on trains (or ships, when it was called *fishyback*). It was only a small step to leave out the trailer.

"It has relations ... and makes a system of them," Michel Serres wrote on the parasite as a constructive element of communicative systems. "It is always mediate and never immediate. It has a relation to the relation, it is related to the related, it sits on the channel."[13] A parasite, because it breaks into a situation, because it partakes in the meal as an uninvited guest, brings new rules. (*Parasite* is from the Greek *parasitos*, "someone who feeds at someone else's table"; *para* means "next to," and *sitos* means "food"). It expands the situation to a system.

According to Serres, the appearance of a parasite always brings about a disruption. The attempt to make the logistical chain a reality without the container as a unifying medium led to broken transit. Simply thinking in the chain, as was done since the 1920s, was not enough to produce a systemic coherence. With conventional transport and handling techniques, the chain revealed nothing more than the sum of its costs and the problems with its links. It was only with the container that the broken transit was interrupted. "What were previously breaking points in combined transit became connections through the container."[14] This is how economic historian Helmut Braun and business economists Robert Obermaier and Felix Müller put it in their joint investigation of the historical role of the container from a logistical perspective.

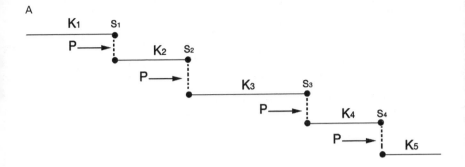

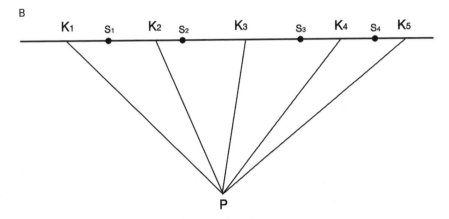

The parasite model of Michel Serres applied to the container system: (a) state of broken transit before the intervention of the container parasite (P), and a chain of separate transport channels (K); (b) reorganization of transport conditions between the K channels under the influence of P. Numerous means of transport (regional channels) (K) and numerous (local) stations (S) are given, at which the channels end and collide, without being directly connected to one another. The container (P) inserts itself between the regional channels and establishes permanent local channels. In so doing, it unites the regional channels into a global channel. The material, visible components of this parasitic usurpation are the container itself and the new parts of the transport system that it carries with it, like container bridges, van carriers or twist locks. The conceptual, invisible components, the thinking that requires the "relation to the relation," is intermodalism. © Alexander Klose.

It interrupts the interruption in the sense that it contains what the others had previously shared and delivered to one another. It connects the transporters more closely because it takes the cargo off their plate. At the same time, it interrupts and ultimately ends a tradition of cargo handling thousands of years old, which got by with few technical aids and required many hands. The container, by insinuating itself as a parasite precisely at the interruptions and bridging the flawed connection points, makes the many channels into one—a channel of containers or a pneumatic tube system. The container has "a relation to the relation" of the means of transport. It is the meta-container and the meta-transporter.[15]

Medium

Even the earliest texts of transportation science speak of the container as a "medium" of transport. Nevertheless, the view of media widespread in many areas of scholarship reduces their meaning to the transmission (or transport), the processing (or handling), and the storage (or warehousing) of information. In contrast, today's media theory involves instances of the in-between, of connection that produces reality in a specific way. There are also historical reasons that it draws no principal differences between the transmission and processing of information, on the one hand, and that of goods or people, on the other hand, since these spheres developed together historically and influence and overlap each other to this day in many ways.

Canadian economic historian Harold Innis and his student Marshall McLuhan, who count among the founders of modern media theory, emphasized in their analysis that communications systems regulate the trafficking of news as well as of goods and people and that they consequently concern a plethora of media, from the road system to shipping to language. Building on this, Friedrich Kittler, a doyen of late twentieth-century media theory (who died in 2011), discussed a "triad of 'things communicated'—information, persons, goods."[16]

Thus, for instance, as early as the great river cultures of antiquity—along the Nile, between the Euphrates and the Tigris, and along the Yangtze—the water current was the medium for the transfer not only of goods and people (i.e., workers) but also of knowledge and of commands regarding water allocation and the levying of goods. Thus, on the one hand, the prerequisite for this complex business of commodities and administration was regulated shipping, and, on the other, "skillfully crafted and optimized transportable writing surfaces," in order to save knowledge and transmit it from one place to another.[17]

To give another prominent example, the rise of the railway system in the nineteenth century is inseparable from the information system of telegraphy. It was only this that made it possible to coordinate travel times in a long-distance transport network characterized by dozens of distinct local times and a general lack of adjustment to the new technical pace. Similarly, the development of steamship travel, and above all of air traffic, is joined at the hip to radio and radar technologies. And today's container logistics would be impossible without computers and the Internet, both to coordinate the complex processes of physical transport and to transmit worldwide the cargo "papers" unified by the "chain."

On both sides, for the transmission of information and the transport of material goods and people, it is about cultural practices that are bound to specific materialities (the "hardware") and that can be characterized by a specific form of organization (the "software"). (In this context, it is interesting to note that the journal *Rationeller Transport* spoke about the organization of transport as software as early as 1971).

Pallets

When containers first came into the focus of the broader public in the 1950s and 1960s, they were seen as chic, modern objects. "The container has the luster of the unfamiliar," an article in the German magazine *Rationeller Transport* said in 1969. Nevertheless, it was a much smaller, unspectacular loading medium that first allowed transport from the factory floor to the final buyer to become a reality: the pallet.

In light of the high technology of the twentieth century, the pallet seems archaic—it's just a wooden platform nailed together with a support frame. However, in combination with motorized methods of conveyance, it is highly efficient. The requirement for its systematic use was met in the 1930s with the introduction of the forklift, an American invention. In principle, the pallet is a transport container reduced to two dimensions. When it is properly loaded, its footprint determines the length and width of the loading volume; the height alone is variable.

In the enormous logistical operations of World War II, the Allied troops were able to reduce the costs for the loading and unloading of goods—whether foodstuffs, weapons, machine parts, or munitions—to less than one-third of what they were with the aid of pallets.

After World War II, the new technology also quickly spread to Germany and other (Western) European nations. The International Union of Railways and the BIC strove for the mutual coordination of the development in the various countries. This led

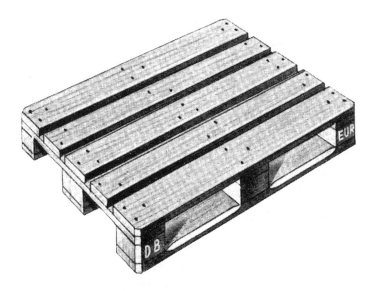

Paletten
Pool-Paletten DIN 15.146/2
für internationalen TAUSCHVERKEHR gütegesichert

Advertisement for the European Pool Pallet, 1960s. In 1964, following the introduction of the pool pallet, German business economist and container specialist Walter Meyercordt wrote euphorically, "All external, intercompany and internal transports, handling processes and storage measures with standardized, interchangeable pallets can be summarized with the proposition CARGO UNIT = TRANSPORT UNIT = STORAGE UNIT. And through the use of consistently standardized pallets an inevitable order emerges— regardless of the stage of production or distribution!" From Walter Meyercordt, *Behälter und Paletten: Flurfördermittel, Lager- und Betriebseinrichtungen* [Containers and pallets: loading devices, storage and enterprise facilities], 2nd ed. (Darmstadt, Germany: Hestra, 1964), 20f.

to the founding of the German Pallet Pool on January 1, 1960, and the European Pallet Pool in the summer of 1961; this was initially joined by Switzerland, Austria, Italy, Belgium, France, Luxembourg, and the Netherlands, in addition to the Federal Republic of Germany.

This was the birth of the Europallet, with the standardized dimensions of four feet by two and a half feet and the quality of the material and workmanship guaranteed by a seal of approval. The businesses taking part in the Pallet Pool had to contribute a certain number of pallets in the system and agree upon monthly exchange rates. Then the pallets could circulate among the firms and the nations on rail and road without consideration of tare rates (i.e., the weight of the empty pallet), with the guarantee that each company would always have available to it the number of pallets that they contributed.

Despite all the ideal models from the early age of pallets and containers, the use of Europallets in international container transport presents problems to this day because they are not adjusted to the normalized ISO containers that appeared five years later. The inside width of the container is a few inches too small to load Europallets perfectly. In this way, up to 30 percent of loading volume is lost with each shipment. Consequently, the Europallet has largely remained a European matter, and a container system adapted to the dimensions of the Europallet has held its own to this day against the ISO container for land transport. In contrast, it was primarily pallet dimensions adapted to the measurement of the ISO container that came to be used in North America and in intercontinental transport.

Theory and Practice

Although the clamor for seamless, global container transport networks had already begun in texts before the first boxes were even loaded onto ships, and even though this clamor has grown tremendously to this day, there has never been a single de facto container transport system, only several in parallel. And all standardizations came in the face of great resistance and at a high price.

The fact that the development into a seamless transport network was anything but smooth is demonstrated by the story of U.S. military transports, which were often viewed as the cradle of modern logistics and the origin of containerization and which should consequently be treated here as a brief example. Indeed, during World War II, the U.S. Army was already using pallets with great success, and a short time later, it developed a transport container that is continually cited as a precursor to today's container system: the ConEx box.

However, with a height of nearly eight feet and a length and width of 6 feet by 6 feet, the box was not much more than a pallet outfitted with walls that was transported with general cargo on classic freight ships. The U.S. Army never succeeded in erecting a true container system network, with special loading machines and box volumes optimized for various transporters and with the subsequent gains in loading capacity and handling speed.

The American military opted to rely on RoRo ("Roll on, Roll off") ships and specialized ship varieties like landing ship tanks, rather than containerization. Indeed, with these one could land in a flat sandy cove without port facilities and offload cargo. However, they transported only a fraction of the cargo of a regular freighter, not to mention modern container ships. The result of this decision was that considerably more ships had to sail.

"The military had the ideas, but they were not prepared to implement them in the system," transport scientist and U.S. States Merchant Marine Academy instructor Gerhardt Muller assured me in May 2005. "Commercial shipping did it."

Thus, it was the Sea-Land Shipping Company that overtook a portion of supply shipments for Vietnam in 1967, ensuring the continuation of this materially intensive war, which had been threatening to end in a logistical disaster. It was a dubious undertaking, from a humanistic perspective, but an impressive demonstration of the superiority of container logistics. With only seven container ships in liner service, Sea-Land reliably delivered 10 percent of the total supplies—above all, food, drugs, and other sensitive goods. For the rest the U.S. Army required more than 250 ships.

Although civilian shipping, and gradually the entire cargo transport industry, shifted completely to containers, with the effects described, the U.S. Army continues to maintain a non-containerized transport fleet. Even in the Persian Gulf War in 1991, all munitions were transported under general cargo procedures. Businesses on the open market took on the "progressive" portion, a situation that has caused considerable headaches for Americans, since the vast deregulation of the 1980s has seen global cargo transport fall almost entirely into the hands of non-American businesses.

Muller sees psychological reasons at play for the fact that it proved difficult for a strictly authoritarian structure like the army to consistently implement logistical principles. Since this would result in restructuring in all areas, particularly in the traditionally more highly valued fighting units, suggestions for reform always failed amid the resistance of high-ranking officers of the navy and the army (who were in a permanent struggle for authority and status, which is also not exactly conducive to a holistic thinking in transport chains). Add to this the fact that as a quasi-monopolist, an army like that of the United States, a nation whose supremacy is based largely on military superiority, does not have to face anything near the cost pressures of a private corporation.

However, even in the market economy, not everything runs according to the optimized-flow ideal. Rather, logistical reorganization causes new inefficiencies. This is shown in an study by Ulrich Welke, a sociologist of work who has dealt with the interplay between the land and water portions of the logistical chain. At various points, he came to the conclusion that since the planning of transport takes place entirely on land, this creates problems and dangers for shipping. The chartering of ships is often organized by nonexperts in large logistics centers who can certainly manage the appropriate computer programs but who are not sufficiently initiated in the high art of trimming, the optimal distribution of weight on a ship. This leads to a situation where ship crews must anticipate and avoid errors that could have fatal consequences for them.

This is also the case when the loading gangs responsible for the fixing of the containers stacked on board do not do their work carefully, as is relatively common in many ports, according to Welke. Because of the extremely high time pressure for container ships in the logistical chain, there is no way to demand exhaustive enforcement of loading operations. After putting out to sea, the ship's crew must then inspect everything as quickly as possible and personally remedy broken lashings, the additional fortifications for the containers stacked on deck through metal clasps. Welke came to the following conclusion:

> Logistics centers don't seem to realize that more goes into the actual process of transport than the average operating speeds on land and sea.... Because in reality, people do not only act (and do not act consistently) as executors of the logistical system. This produces friction. Sailors can fight back the least. Consequently, they are forced to overcome the frictions of container transit through additional effort.[18]

Ultimately, logistics as well is a normative ideal, no different from rationalization, which always fails amid the stubbornness of existing structures: people, materials, technologies, and organizational forms. Fortunately so, one might add, for how could one find a foothold if the world really became completely fluid?

The Power of a Third Party

The development of container shipping as well as that of the entire logistics business is largely determined by newcomers and lateral entrants to the field and their unorthodox business ideas. Malcom McLean's Sea-Land Company largely attained its success despite the traditional shipping companies and around governmental regulations. The same is true for the Danish Maersk Line. Indeed, A. P. Møller, which owns Maersk, is Denmark's state corporation, so to speak, which has transport rights for Danish maritime oil and gas resources and also owns the largest supermarket chain.

However, the one man, one boat enterprise founded by Captain Peter Maersk Møller in 1904 owes its ascent to a consistent search for and occupation of niches and for business policies beyond the price agreements among the large international shipping companies that regulated the ocean transport business into the 1980s (and likewise beyond the large consortia and charter agreements with which the traditional corporations saved themselves in hard times).

Since the inception of its container ship line in 1975, Maersk's operation was emphatically, even arrogantly, stubborn. And it grew and grew with this strategy. Through the acquisition of Sea-Land (which belonged at the time to the American railway consortium CSX, as mentioned in chapter 3), it moved to the top of the list of the world's largest container shipping lines, and it has expanded its lead ever since. To this day, it is notorious, even infamous, for its maverick, secretive business policies.

In second place on the current list of the largest container shipping lines is the Mediterranean Shipping Company (MSC). Founded in 1970, the company is based in Geneva—not exactly known as a traditional sea trade port. It established its rise with *connecting carrier agreements*, contracts with different shipping lines for the transport of containers to ports that it did not service itself. In contrast to the usual *space charter agreements* executed between the large lines, these contracts are concerned only with the actual transport freight and not with a right to space that must be paid regardless of whether it is filled, as was and continues to be common among the large shipping consortia in order to have a measure of certainty in planning.

The most important contractual partners were the NVOs, a new category of logistics providers created in the 1960s that organized intermodal door-to-door transport and took on, above all, the administrative work. In other words, they issued freight documents and assumed guarantees for time and place of delivery without having their own means of transport.

Thus, for a long time it was possible for the MSC to underbid the rates of its large competitors in intercontinental liner service because the company operated exclusively with secondhand ships (like the unfortunate MSC *Napoli*, with which this book began). And since it kept the shore-based staff as small as possible, the MSC expanded above all in niche markets, such as with a line between the western coast of South America and the East Coast of the United States. Only at the end of the 1990s did the MSC acquire its own ships. Amid the further consolidation of the international shipping industry around 2000, it moved past all traditional carriers into the upper echelon.

Another company that must be named in this context is the Taiwanese shipper Evergreen. Established in 1968, its early business success was also based on niche offerings and lower rates than those of the corporations in the "liner conferences." These were agreements on routes, rates, and schedules that were made between the dominant shipping companies to avoid the risks of fierce competition. The company expanded enormously in the 1980s and became the world's largest container transport firm for a few years, since it was able to establish itself despite McLean's U.S. Lines, with its global liner transport amid the crisis of the 1980s, thanks to faster, smaller ships and lower costs (as described in chapter 3).

Wherever one looks in the diversified economic events of transport logistics, the power of third parties is apparent. This is also reflected in the common umbrella term for this kind of business: third-party logistics, or 3PL.

Operating complementary to the NVOs are the tramp shipping companies. These are transporters that offer up their ships to other transport companies, charterers, or liner services for a brokered rate. Only one-third to two-thirds of the large liner shipping companies' vessels are their own. The rest are chartered in order to minimize the risk of ships not working to capacity in the event of temporarily low transport volumes. The typical tramp transporter is at least as much a financial juggler as a shipping operator. This conjures up memories of McLean and his financial skill. Thanks to tramp shippers—like Rickmers, NSB, or Claus-Peter Offen—and investment-friendly fiscal policy, Germany is currently the world's leading nation for shipping, measured by tonnage in German possession, and Hamburg is the most important place for the financing of shipping corporations.

Another area of the container business where the power of third parties can hardly be overlooked is the box itself. Since there is no exchange agreement among all large container transporters comparable to the European Pallet Pool, the market and organization of boxes is rather complicated. Bottlenecks continually occur because not enough containers are available. The question of how to deal with empty containers is particularly a problem, given the unequal volume of freight moving from and to East Asian ports.

Consequently, empty-container logistics—stockpiling, maintenance, and disposal—has established itself as an independent branch within logistics, taken on by the large port operators and shippers, on the one hand, and by third-party firms, on the other. The container stacks in the foothills of the harbor regions, as high as towers and as broad as houses, stand as evidence of this fact, visible from miles away.

The containers, a significant purchase investment on their own—which is certainly the decisive difference from pallets—belong only in part to the container transporters. Today, the production of a 20-foot container costs around $2,000, and a refrigerated container costs around 10 times as much. Over three million TEUs were produced worldwide from 2006 to 2008. As a result of the recession brought on by the global financial crisis of 2008, this number went significantly down in 2009 and 2010. But since 2011, the number is above 3 million again.

Profiting from the situation are, on the one hand, Chinese companies, which currently produce more than 90 percent of all containers worldwide. With a share of more than 50 percent, the market leader is the CIMC group, based in Shenzhen. On the other hand are the container leasing companies, to which approximately half of all containers worldwide belong. This business started in the United States in the mid-1960s, and American firms have dominated to the present day. The largest European container renter is the Hamburg company Capital Lease, recognizable by the dark green container with a white label that could as easily be on the cover of a business magazine.

Container leasing companies maintain global networks of container depots. As a rule, they not only lend containers, they also assume responsibility for their maintenance. Sometimes they offer tracking systems or rolling material like truck chassis or railway flatcars, or they take on the entirety of container logistics. In this respect, they are classic representatives of 3PL, so-called integrators or contract logisticians that act as third parties, taking over planning, storage, distribution, communication, or customer service from other firms. The container depots are somewhat like distribution centers for the boxes themselves.

If one considers the booming industry of logistics, which has grown to become the third-largest sector of the economy, one sees an extremely complex web. There is hardly a company that can be described by a simple product profile. Classic carriers like the German company Dachser, which began with cheese transport from Allgau into the Rhineland, offer contract logistics services as well as traditional shipping and railway ventures. The fragmentation, abstraction, and mixture of the production and distribution processes is accompanied by confusing proliferations, bizarre specializations, and hybridizations of companies and undertakings.

That is the new logistical order: operational chains that tangle into complicated graph networks. At every table, there is always a third, a fourth, and a fifth. And the table is the size of the world.

Cybernetics 2.0: Robot and data container for (fictional) research use on Mars. © Sun Microsystems, https://photos.sun.com. Used with permission from Oracle.

6 Computing with Containers

Hiro spends a lot of time in the Metaverse. It beats the shit out of the U-Stor-It.

—Neal Stephenson

In October 2006, Sun Microsystems, one of the world's leading developers and suppliers of computer systems, proposed Project Blackbox: mobile, modular data centers in 20-foot ISO containers. According to the company's reasoning, in a time of exponentially growing demand for computing capacity that often appears on short notice in unexpected places, preinstalled, largely self-supporting, quickly deployable, and easily expandable modules are the right answer.

"Project Blackbox is a build-once, deploy-anywhere, modular data-center that delivers flexibility to locate computing resources where you want them, when you want them—anywhere in the world." This is the description in an advertising brochure that was available from the Sun Microsystems website in the initial phase of the project in 2008. Areas of application for this new kind of modularized data center, which has since been offered by a whole array of firms, include not only crisis situations like wars or natural catastrophes but also the entirely normal expansion or even new establishment of data centers for large corporations.

The computer networks within each data center are designed to be easily combined with other data center containers into large, expandable networks. In addition, because of new cooling and packing techniques, vastly more computers fit inside the container than in the same space in a conventional data center, and much less energy is required for its cooling.

More flexible allocation of capacities through modularization and standardization, consequential (continued) networking, externalization, and distribution of sites of production (or processing)—the design follows the same pattern as the transformation of cargo transport logistics and the resulting fundamental changes in the production of goods in the last 50 years brought about by containerization. As the black box becomes physical reality, a container filled with computers is nearly emblematic for a situation that is often gladly overlooked in much of the modern discourse on so-called globalization. When attempting to find explanations for the global coalescence of realms of experience, the image of global digital communication networks comes to the fore. However, the transmission of data is only one side of the "global village" that the pioneer of media theory, Marshall McLuhan, described. The other is the transport of people and products.

In the logistical world of transport and communications networks, material and immaterial elements are intertwined. As the central component of modern logistics, the container does not merely operate as a universal, indifferent transport unit in the heavy space of physical goods. It also functions as a conceptual or symbolic element in the light sphere of electronic data processing. In other words, containers process both physical and data.

As both symbolic and physical media, they work in two directions. On the one hand, they are agents of the digital in the

Computing with Containers

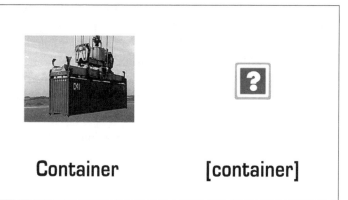

Fluid transition among hardscape, softscape, and imagescape: the container as physical operator in the transport system and as image and conceptual metaphor on the graphical user interface of a computer. © Alexander Klose.

physical spaces of transport, in which everything is subjected to their logic of the processing of standardized, clearly separate (discrete) units, and in this respect the computer-driven control is doubled. On the other hand, they operate as agents of materiality in the symbolic spaces of data processing, where containers implement a specific kind of physical spatiality and where metaphors from the realm of transport (and architecture) play a central role in the construction of the systems.

Computer and container technologies, digital computing with machines, and industrialized container transport developed parallel in the same period and influenced one another. Today's logistics of precisely timed, complex production and distribution processes would be unthinkable without the data processing power of computers and the fast information transfer speeds of the Internet. Container terminals and the organization of logistical chains count among the earliest civilian applications of computers.

Conversely, from the architecture of the computer to the construction of the transfer format for data, the form of digital computer networks, implemented in electronic circuits, largely follows a logic of transport and modularization. The following sections will be concerned with a few of these interrelationships: computing in containers and containers for computing.

Operations Research

The advertisement brochure for Project Blackbox says, "Targeting large-scale network service deployments such as Web 2.0 expansion, field deployment operations, emergency response, and datacenter consolidations and build-outs, Project Blackbox delivers exceptional energy, space, and performance efficiencies. It also provides instant-on expansion and deployment opportunities for any organization wanting to move away from the rigidity of legacy datacenters in pursuit of maximum savings and operational flexibility."

Data center containers are deployed—positioned or installed in an operation—regardless of whether the operation has to do to with the mathematical coordination of a satellite-assisted air-to-ground attack, aiding in the reactivation of a power plant damaged by natural forces, or expanding the computer capacity of a university library. Although the word *deployment* is taken for granted in business speech and is generally used without the least bit of militaristic ulterior motives, to *deploy* maintains a primarily military semantic context, as the dictionary reveals. It means to position in readiness for combat, to bring into action, to place in battle formation, to spread out, and to distribute systematically or strategically.

Computing and communications containers for mobile deployment in the field by the U.S. Army have existed longer than comparable products in civilian industries. However, the mere fact of the use or origin of a certain technology in the military is perhaps less meaningful than it seems at first glance. For instance, as has been shown, the pioneering role of the U.S. Army in the conception of container transport systems had no lasting influence on their further development. The ConEx boxes rusted in their areas of operation in Southeast Asia, whereas a container system developed by civilian companies broke into the market economy and ultimately caused the army to reform its logistical concepts.

In this case, it seems that a certain thinking in actions, operations, and systemic concepts—a thinking in logistics—was more decisive than actual material technical transformations. With its terms and methods, this brought a specific military heritage into the economy whose continued existence, in the words of Viennese philosopher and media scholar Wolfgang Pircher, "sparked suspicion that, since World War II, the state of war has become permanent."[1]

The approach of operations research, developed during World War II, may be seen as the main channel through which military-logistical thought has entered the organization of civilian society since the 1950s. Essentially a mathematical method for the interpretation of war-related data that was of great importance in the Allied battle against German air attacks and German U-boats and that gained a certain fame in the postwar years, operations research quickly became one of the most sought-after tools in the economic efforts at rationalization through the optimization of business processes.

According to the methods of operations research, optimizing a system meant finding the best possible combination of elements rather than trying to maximize the performance of each individual element. To achieve this, optimization strategies analyzed the interplay of all elements and then concentrated on the bottlenecks, the elements that curbed the overall achievement of the system.

On the one hand, the use of new information technologies had played a key role since the earliest days in anti-U-boat warfare and the development of radar systems, because the computing operations were too complex and especially too extensive to be done without the help of a machine. On the other hand, with its aura of war-deciding information technology, operations research served to open the door for the still extremely expensive and extremely complex computer technology in areas of civilian administration and private enterprise.

After the war, in 1946, RAND (short for Research and Development) was founded as a joint venture of the U.S. Army Air Forces and the armor company Douglas Aircraft. RAND became the most important agency of operations research, practically the embodiment of its spirit.

Recast in 1948 as a private nonprofit research organization based in California, the institute, now called the Rand Corporation, initially had the primary aim of probing technologies for air war. However, since the 1950s, other strategically important institutions, like the U.S. Atomic Energy Commission, have numbered among its sponsors and clients. Situated in the midst of academic, industrial, and military research, the system-based approach fostered by the Rand Corporation significantly shaped U.S. foreign and military policy during the Cold War. At the height of this concept, in the mid- to late 1950s, the institute employed more than 2,500 staff members. This included some of the most renowned mathematicians, physicists, and engineers of the period as well as social scientists and economists.

The Rand Corporation was a center for civilian intellectual participation in military-strategic and technical problems. The research done there brought about a transfer of knowledge in these areas. Military-strategic calculations were charged with fundamental theory and social science, and the handling of economic and social organization was encoded in military practice (logistics). This was ensured by a steady transfer of players from the military into the civilian realm and vice versa.

Among these border crossers was Austrian economist Oskar Morgenstern, who emigrated to the United States in 1938. Together with mathematician John von Neumann (who was also a computer pioneer and a key figure in the construction of the first atom bomb) he founded mathematical game theory. In 1950, Morgenstern, who was then both a professor at Princeton University and a researcher at the independent Institute for Advanced Study in Princeton, New Jersey, gave a lecture entitled "Notes on the Formulation of the Theory of Logistics" at a logistics conference supported by the Rand Corporation. (Five years later, this text would appear as an essay in the pertinent journal *Naval Research Logistics Quarterly*.)

This text may be seen as the first in which the principles of military logistics were transferred to the realm of civilian economics and in which both were subjected to a systematic comparison. He came to the conclusion that on the level of individual operations, military logistics was concerned with incomparably larger quantities and a multitude of uncertainties and rapid changes in fundamental premises, and in this sense it was the more complex process, as the military master plan must always be kept in mind.

In contrast, the logistical operations in the civilian world were spread across a plethora of comparatively manageable submarkets. However, if one considers the complete context of all undertakings involved in production and distribution processes, Morganstern concludes, the civilian economy is capable of dealing with much greater complexities than military logistics, provided that there is systemic thought and that optimization strategies are applied in logistical chains. According to Morganstern:

> The simplicity of business logistics derives not only from the much smaller numbers of items even for the largest firms compared with, say, those encountered at a single Naval supply depot, but also from the fact that there is, as a rule, an instantly accessible source available in the market. Instead of having to provide a vast closed logistical system, it is possible to restrict oneself to a much smaller individual one and to supplement it as needed from the market. The overall inclusive individual military plan is replaced by a composite organization of many firms communicating with each other rapidly and accurately through the open market and the price system.[2]

Added to this possibility—that of an accumulative approach through restriction to individual segments and of comparatively unproblematic access to required materials—is the fact that it is distinctly easier in the free economy to come by the applicable data for all phases of the processes, in order to optimize them in the course of a broader systematic analysis. Here, however, the relationship between military and economic logistics is reversed. By virtue of its greater transparency and predictability, the latter makes considerably larger gains in efficiency possible. It is only in the civilian economy that the use of a few central components of scientific warfare, particularly its logistical principles, associated above all with operations research, are fully developed.

The fact that this situation remains unchanged to this day, that a military way of thinking can still achieve its greatest successes in the market economy, is demonstrated by the case of the American Lieutenant General William Pagonis. During the Persian Gulf War in 1991, this man was responsible for all military logistics of the U.S. invasion to liberate Kuwait, in the course of which more than 1 million tons of goods were transported in more than 40,000 containers to Saudi Arabia and onward within six months.

However, although the entire operation ran according to the most modern logistical practices, voices were subsequently raised that complained about a gaping hole between the claims and the reality of military just-in-time logistics. For instance, some American units had to get by for days without food, whereas others were reportedly instructed to buy water from Iraqi merchants.

Two years later, Pagonis became the vice-president of Sears, Roebuck and Company and thus became the man responsible for the logistics of an ailing firm that had posted record losses of $3.9 billion in 1992. Thanks to the reorganization of the entire logistical operation, within a short period, the company (now the second-largest retailer in the United States) realized savings of more than $1 billion a year—one of the most successful restructurings in U.S. economic history.

A similarly positioned example operating within the same industry is the strict logistical organization of the world market leader Wal-Mart, whose militaristic business style and rigidly hierarchical decision-making structure has been reported (and lamented) by many insiders. For instance, a manager who oversaw the direction of a Wal-Mart distribution center in the early years of this century as an employee of the contract logistics firm Exel had the following to say in an interview about Wal-Mart's business style: "I had to deal with eight of their managers, who took turns giving me grief.... The main person was ex-military, out of West Point. There are lots of West Pointers in Wal-Mart's management. They operate with a strict chain of command."[3]

Basic Systematic Research in the Early History of the Container System

What does all of this have to do with computers and containers? After World War II, banking was the first nonmilitary sector of the economy in which computer systems were systematically introduced, albeit mostly without the desired gains in efficiency. Computers and computer-assisted basic research also played a role in the development and implementation of container logistics that should not be underestimated. Economist Joshua S. Gans wrote in a 1995 discussion paper for the Kensington School of Economics that "containerization requires far more than its predecessor in the way of detailed information regarding cargo types, ownership, and destination so as to assign proper containers and to keep track of the contents of those containers.... The development of operations research and then computers were crucial ingredients in the success of containerization and aided in its effective spread."[4]

If we wish to explain how the decisive innovations in the transport industry and the rise of container logistics came about, then we must consider another man, and an enterprise with an extremely different strategy, beyond the much-cited heroic story of Malcom McLean and his company Sea-Land (see chapter 3): Foster S. Weldon and the U.S. West Coast shipping company Matson.

The Matson Navigation Company is a firm with a long history and roots in the nineteenth century, widely known during the first half of the twentieth century for its passenger and freight shipping from San Francisco, Seattle, and Los Angeles to Hawaii and back. Although the company was securely positioned as a niche operator, in the early 1950s Matson's management began to consider possible rationalization measures in light of increasing dockworker compensation and stagnating productivity. To this end, Stanley Powell Jr., later the president of the company, was chosen to be leader of a research team in 1953 that was initially a one-man undertaking.

Powell came to the conclusion that a lack of standardized cargoes and processing equipment was to blame for the problems of the shipping company, and he pleaded for a restructuring of the transport processes along the lines of the model of Ford's automobile production. In 1956, this led to the founding of a separate research division—the first of its kind in the shipping industry—under the direction of Foster Weldon, a nonspecialist operations research expert and a geophysicist who had carried out 15 years of fundamental research for weapons systems on behalf of the army (he was ultimately involved in the development of a nuclear submarine at Johns Hopkins University in Baltimore). Weldon approached the subject of cargo transport systematically.

In a certain sense, he was the precise opposite of the informal trial-and-error style of McLean and his team on the East Coast. Here was a scientist formally educated at the best institutions who was connected to the scientific community, who published the results of his research, and who used advanced scientific research methods and equipment. McLean was a self-made man without a degree who surrounded himself with pragmatists and revealed his solutions only at the last moment, with as much spectacle as possible.

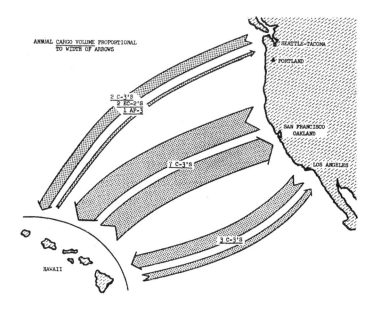

The inventory and description of existing conditions as a systemic network is Weldon's point of departure for the systematic probing of the foundations of a future container system: "To provide a yardstick with which to measure the changes induced by containerization we must first define the physical facilities, the cargoes, the organization, and the costs that comprise the present … transportation system." Reprinted with permission from Foster L. Weldon, "Cargo Containerization in the West Coast–Hawaiian Trade," *Operations Research* 6, no. 5 (Sept.–Oct. 1958). Copyright 1958, the Institute for Operations Research and the Management Sciences, 5521 Research Park Drive, Suite 200, Catonsville, MD 21228.

Whereas those responsible at Sea-Land negotiated contracts for future transport business without knowing how they would accomplish this technically, relying instead on intuition, luck, and chutzpah, Matson undertook comprehensive series of tests and simulations under the guidance of Weldon, in order to draft an optimal system of containers, vehicles, ships, and processing equipment for the company's transport.

The researchers developed a complex simulation model that incorporated an entire year's worth of data on the volumes and costs of more than 300 categories or goods for each port that the company served. Added to this were the costs for dockworkers, use of the dock, and handling equipment capacities; the precise composition of the cargo for each individual ship; and costs for temporary storage, transfer, and truck transport inland.

On the basis of these data, the researchers simulated scenarios with different cargoes, volumes of containers, types of loading, and shipping routes. They looked for answers to the following questions: What is the optimal container size? Should large ships just commute back and forth between large ports and give their freight to smaller feeder ships once there (known as the *hub-and-spoke system*), or is it more efficient to include multiple smaller ports in round-trips? At what time should a ship set out from Honolulu in order to minimize total costs for shipping a load of pineapples to Oakland?

In the 1950s, such simulations, whose origin lay in the scientific warfare of World War II, were an absolute novelty beyond the military. No one in the transport industry had ever tried such a method. The analysis of thousands of shipments was accomplished through punch cards and rented space on an IBM-704 computer. At this time, when computers were still large and extremely expensive machines that were produced only in small

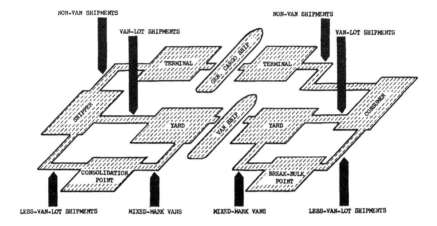

Notwithstanding the idealizing model of container transport from door to door that exclusively involves full containers, in the reality of goods transport between the U.S. mainland and Hawaii, one must also allow for goods that cannot be containerized and for a handling procedure for quantities of goods that fill up less than a container, in order to arrive at a realistic model for calculations. "In practice, the concept of van [container] operations is not quite as simple as the ideal case.... In the first place, there are several classes of van shipments in addition to full van-lot shipment destined for a single consignee. Then too, van shipments are only a part of the cargo to be processed. Overall, the total organization of the system must comprehend a variety of operations." Reprinted with permission from Foster L. Weldon, "Cargo Containerization in the West Coast–Hawaiian Trade," *Operations Research* 6, no. 5 (Sept.–Oct. 1958). Copyright 1958, the Institute for Operations Research and the Management Sciences, 5521 Research Park Drive, Suite 200, Catonsville, MD 21228.

quantities, this was the usual way of gaining maximum access to computing capacity, and it also ensured that the few machines were almost completely utilized.

In February 1958, Weldon submitted a kind of manifesto of container logistics to the journal *Operations Research* with the results of his two years of research. The essay was published at a time when Matson's container program was already in full swing. He began with the basic observation:

> The containerization of general-merchandise cargoes is under active investigation by almost every major railroad and steamship operator in the country.... In spite of all these efforts, however, there are no clear guidelines for the design of optimal containerization systems.... All transportation companies have their own pet theories on the detailed equipment requirements comp[o]sing a "best" container system, but there are no quantitative data relating even such gross characteristics as container size to the economics of a total transportation operation.[5]

Among other things, the study revealed that of the container dimensions tested (12, 17, 24, 35, and 40 feet long, 8 feet wide, and 8.5 feet high), a 24-foot container would be the ideal unit size for standardized container transport between the West Coast and Hawaii and that specialized cranes should be affixed to the docks rather than using cranes mounted on the sides of ships, as had been done by Sea-Land in its early years. The specific circumstances of a situation not oriented toward expansion (at least for the time being) led to this result. In addition, the considerable spatial limitation on the islands and in several West Coast ports made the stackability of containers seem advisable. In the long view, these decisions made by Matson proved to be more enduring than the ad hoc solutions of Sea-Land, which remain among the basic foundations of the container system today.

This also gave rise to the prototype of the container bridge, which would become the worldwide standard, under the aegis of

Matson's chief engineer in collaboration with the Pacific Coast Engineering Company (Paceco). For years, Paceco's cranes with the A-shaped frame left their mark on the image of containerized dock handling and at times nearly became a synonym for the container bridge itself.

However, from today's point of view, the most durable result of the Matson-Weldon study may be the constitutive distinction between a *full container load* (VL = van-lot; Weldon, like McLean, took a truck body as starting point for container dimensions), a load that fills a container, and a *less than container load* (LVL = less-van-lot), a load that makes up less than the capacity of a container.

Unequal cargo volumes in European and U.S. transport from and to east Asia present a grave problem today. The history of containerization is fundamentally shaped by the factoring in of growth, by the effects of an *economy of scale*—that is, one attempting to compensate for possible losses from empty runs by further increasing the load capacities for the full runs, as well as by the relative minimization of costs for the individual boxes compared to the total volume.

In contrast, Weldon proceeded from a limited situation that he attempted to optimize through his mathematical calculations. Through these, he came to the conclusion that despite the additional loading costs, it was cheaper to repack containers that were not completely full at the port of departure than to send the less –van-lot all the way. In order to avoid having to repack these mixed loads once again, the containers were packed in the order of delivery. In this way, trucks with full containers can set out directly from their destinations on a round-trip to the various targets and gradually deliver their cargo.

Computing with Containers 215

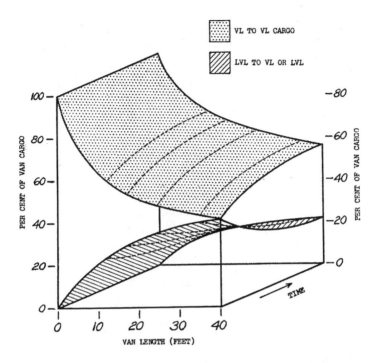

The relationship of full container loads (noted here as VL) and loads with less than container volume (noted here as LVL) in the comparison of different container dimensions. A 24-foot container yields the best relationship. Depiction of the results of computer-assisted analysis by the Matson research division. Reprinted with permission from Foster L. Weldon, "Cargo Containerization in the West Coast–Hawaiian Trade," *Operations Research* 6, no. 5 (Sept.–Oct. 1958). Copyright 1958, the Institute for Operations Research and the Management Sciences, 5521 Research Park Drive, Suite 200, Catonsville, MD 21228..

Against the backdrop of tremendous rates of increase in freight volumes in container transit, this approach seems comparatively anachronistic and, at most, useful in limited niche situations, as with transport to Hawaii. However, with today's latest technology for the automatic identification of small loading units, with the further differentiation of delivery transport in the course of just-in-time production and the customization of consumer goods, and with the restructuring and further refinement of the services of the transport industry according to the paradigms of parcel and express services, organizational thought in less than container loads, which stood at the origin of containerization and was somewhat overcome by it, gains new currency.

Thus, the operations research approach, which shows the way theoretically for the logistical conception of the container as an operationally closed box, already seems to contain the seed for the demolition and further differentiation of this hermetic system, the path to opening the black box.

Black Box

The black box is a theoretical and practical model for the reduction of complexity. It was introduced in cybernetics as a practical means of dealing with systems "whose internal mechanisms are not fully open to inspection," as pioneering cyberneticist W. Ross Ashby wrote.[6] Like operations research and game theory, cybernetics emerged directly from the military research structures in and after World War II. The core of the idea was that we regularly have to contend with "a certain concealment" of relationships in daily life as well as in research. Rather than inquiring about the inner functioning or even the essence, the cybernetic approach consisted of formulating a set of questions,

Computing with Containers 217

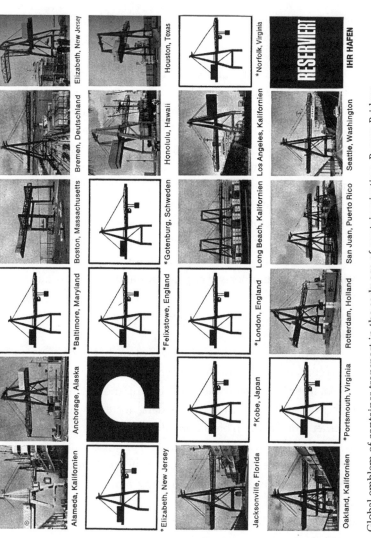

Global emblem of container transport in the early age of containerization. Paceco Bridges, magazine advertisement from 1967. *Hansa: International Maritime Journal* 104 (1967).

as Ashby called it, a protocol according to which one could submit the inaccessible relationships to a kind of investigation and test its reactions.

From these results, one could deduce the composition of the inside of the black box. More decisive, however, is the fact that the question-and-answer framework produces a context of communication and interaction. The framework defined by the questions is the *interface*, the only channel of expression for the context otherwise regarded as a closed container, as a black box, and thus for the system respectively defined by a manageable number of parameters. From its reactions and those of many others, one develops a metasystem into which the individual system contexts are built as reaction apparatus. Ultimately, it amounts to a delegation of responsibility to various levels or various phases in process cycles, when multiple black boxes are instrumentally coupled to each other in a system or brought into a hierarchical order (as is often the case in data trees and folder structures in computers).

The principle of the black box was of central importance in the second half of the twentieth century, both in the construction of technical systems and in the organization of institutional processes—in the organization of data, for example, through Internet protocols or in object-oriented programming languages, as well as in the organization of physical processes, such as serial production in the factory or transport logistics. Each situation in which a black box operates is defined by a limited set of actions and/or communications. Thus, the superior efficiency and speed of the container freight system relative to the methods of traditional cargo transport consists in the fact that the metacontainer appears on various levels of the transport event as a black box. It appears on the level of technical processing, since this can be

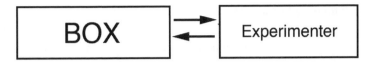

Diagram of the basic black box structure according to Ashby. "To start with, let us make no assumptions at all about the nature of the Box and its contents, which might be something, say, that has just fallen from a Flying Saucer. We assume, though, that the experimenter has certain given resources for acting on it (e.g., prodding it, shining a light on it) and certain given resources for observing its behavior (e.g., photographing it, recording its temperature). By thus acting on the Box, and by allowing the Box to affect him and his recording apparatus, the experimenter is coupling himself to the Box, so that the two together form a system with feedback." From W. Ross Ashby, "The Black Box," in *Introduction to Cybernetics* (New York: Wiley, 1956), p. 87. © Alexander Klose.

restricted to the standardized, procedural handling of invariably identical objects with invariably identical specifications.

The only question that presents itself is whether a container conforms to norms; if not, it must be discarded. Questions along the lines of "Are you heavy or light? Are you delicate or resilient? Are you stackable or will you deform?" don't arise at all, or at most to a small degree. The characteristics of the loaded good are ideally only of concern at the beginning and the end of the transport process and can be largely ignored for its entire duration. The same is true for the organization of data on the transport process. Here as well, a kind of black boxing takes place through the introduction of standardized loading documents and query protocol that allow the individual information on the individual goods being transported to be recalled over the entire length of the transport chain and only in individual, preferably standardized, retrieval situations.

Computing Containers

At its heart, ocean shipping is a network business, just like airlines and telecommunications. Passengers, bulk goods, data all represent uniform-size cargo, shooting through global transport and sorting systems 24/7/365. Viewed this way, airline seats, data packets, and 40-foot shipping containers are much the same: commoditized units for carrying content.

Stewart Taggart brings the structural similarities between container and data transport to the fore with a certain emphasis in an article in the October 1999 issue of the computer magazine *Wired*, entitled "The 20-Ton Packet: Ocean Shipping Is the Biggest Real-Time Data-Streaming Network in the World." These isomorphies, as the cyberneticists call them, have been frequently pointed out in recent years. They form one side of the tight interpenetration of computer and container technologies. Another aspect is their inseparable interconnection through material history.

Computer historian James W. Cortada exposed this relationship clearly. In his large study *The Digital Hand*, he investigated how the introduction of computers in the United States changed the work processes in the various branches of industry. The starting point for his chapter on the transport industry was the finding that in "the second half of the twentieth century, the most fundamental trend in transportation involved its continuous integration backward into manufacturing and forward into retailing processes." The prerequisite for this integration and all of its resulting rationalization effects have been cross-sector information systems: "the glue that held compatibility and integration together from one sector of the economy to another was the collection of information systems."[7]

Consequently, it was this observation that allowed Cortada to understand the principle of intermodality in transit as the merger of various transport systems and information technologies. Thus, the computer may be seen as another third party, along with the container, that breaks into the old order of transport and creates a new one with and alongside the container.

Although a clear separation is not possible, and in fact the individual applications and areas overlap and supplement one another, one can discern two broad levels on which the use of computer technology was and is critically important for modern container logistics as well as for its historical development. It is used, on the one hand, for the control and optimization of physical processes and their systemic construction and, on the other, for the communication of the data associated with these processes. Shipping historian Frank Broeze writes that containerization was

> totally dependent on electronic data processing for virtually every aspect of its operations; one might well say, paradoxically, that computers formed the software of the container system. The management of the container park on a terminal ... was simple compared to the global tracking systems that had to be created to help locate each individual box.... The optimal loading of ships, with boxes destined for a multiplicity of ports and destinations, was an equally useful field for computer application, but as each container could in principle have a different weight and centre of gravity, the ship's stability was of as much importance as was commercial efficiency. Computers soon were indispensable in producing and handling the complex paperwork necessary to document the movements of each container.... More recently, the Internet has become the cutting edge of communications technology for freight bookings. Leading container operators were in the forefront of e-commerce.[8]

As has been shown, the implementation of computers had already begun in the formation phase of the container system in Matson's research division. At the time, other companies still made do in the organization of container terminal processes with large metal boards where magnets were pushed back and forth for each individual container. However, as early as 1958, the large East Coast competitor Sea-Land introduced an IBM-360 computer system in its administrative building on the grounds of the newly constructed container hub Port Elizabeth in New Jersey, the world's first port designed around the demands of container transport, in order to coordinate the movements of ships, containers, and trucks.

And in the pioneering operation in the war zone of Vietnam described in chapter 5, it was an electronic inventory system, still based on punch cards at the time, that assured Sea-Land its convincing success. With it, the cargo and location of each individual container could be tracked from loading in an American harbor to arrival in Vietnam to its return to the United States.

In the late 1960s, a new branch began to form in a supply industry that provided the expanding logistics branch with machines and software. Among the new assignments were the installation, operation, and optimization of technical facilities like terminals, warehouses, and distribution centers, programs for the coordination of complex organizational processes for trucking, rail, train, and shipping companies, as well as port operators and programs for storage planning, for the management of customer data, for the calculation of freight rates and inventory systems, and for the processing and standardized transmission of bills of consignment and all other data in the logistical chain.

The continuous control granted by this system made it possible to shift the focus of business organization from inventory (i.e., stockpiling) to information to the control of current and future processes. It laid the groundwork for the introduction of the just-in-time economics that characterize the modern global production and distribution of goods.

Computing with Containers

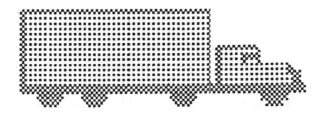

Vision of the intermodal transport system, 1989. Toward the end of the 1980s, no topic was the object of more symposia, forums, seminars, reports, and journal articles than the possible changes stemming from the introduction of electronic data transmission in transport processes. Graphic from the April 1989 cover picture of *Containers*, the official journal of the BIC.

The complexity of processes demanded the implementation of electronic port logistics systems, as was already noted by experts in the early 1970s, such as in an article entitled "EDP System is Key to Door-to-Door-Moves" in the May edition of the trade journal *Container News* in 1970: "If containerization is to fulfill its promise of being highly beneficial to shipper and carrier, it must be backed up—one might even say: it should be led by—computerization." In Germany in 1973, 108 harbor transport companies joined forces and established the Database of Bremen Harbors.

In 1976, the first project of this new merger began—the Computer-Oriented Method for Planning and Process Control in the Seaport (COMPASS). This was thought to be the first cross-company port information system in the world, as stated in a 1991 brochure from the Bremen Warehouse Company. Two years later, in 1978, CT-on-Line, the first computerized system for monitoring and control of the terminal, was installed in the same place. In 1983, it was expanded.

In the 1980s, this kind of centrally controlled port logistics systems was introduced in all of the large ports—for instance, in the summer of 1983, when the Data Communication System (DAKOSY) was adopted in the port of Hamburg. The tentative endpoint of this development was fully automated, computer-controlled container terminals like the first of its kind, the Europe Container Terminal in the port of Rotterdam, or the Container Terminal Altenwerder in Hamburg, brought online in 2004.

Still, in 1973, Wolfgang Bohle, author of the Bureau International des Containers (BIC) organ *Containers*, complained that goods transport proceeded more quickly than data transport: "At present, the shipping companies are continually confronted with the fact that the containers stand ready for outgoing shipment, ...

but ... the papers necessary for processing are lacking." He came to the conclusion "that the future task of container transport no longer lies in solving technical questions ... but rather ensuring the simplification of documentation and thereby its acceleration."[9]

This situation changed fundamentally with the introduction of the Internet at the beginning of the 1990s. This marked the change from centrally guided to distributed information and control systems. The online-supported organization of container transport makes the most varied business models possible—one-to-one, one-to-many, and many-to-many—through electronic data interchange between in-house computer systems, Internet forms, and email.

Since the late 1990s, with decentralized networking through the Internet, third parties, called *infomediaries*, have come into play as a new subspecies of the nonvessel operators discussed in chapter 5. They offer to handle the organization of the entire flow of information between transporters and loaders on their own platforms, and they also engage somewhat in the business of short-term chartering of container spaces on ships.

The large shipping corporations initially attempted to oust the new third parties from the market. When that failed, they formed coalitions. Thus, the two largest infomediaries, GT Nexus and INTRA, cooperated with nearly all the world's important container shippers, or the latter decided to completely outsource their own information technology (IT) departments and enter the market as infomediaries themselves, like the shipper OOCL with CargoSmart.

The technical implementation of intelligence in transport chains within electronic communications systems certainly supported cross-industry networking since the 1970s. However, because of technical limitations, it remained far behind the claims of a radical transformation of the container transport concept. Complementary to the metastructure of the container system, the metastructure of the Internet created the conditions to let the specific qualities of the individual carriers recede further into the background, in order to create at least the illusion of a seamless transport system, as has been said in numerous publications in recent years.

The relationship between the transport of goods and that of their data has been reversed. The promise of (supposedly) massless and unopposed movement contained in the data transfer being carried out at light speed through the Internet became a paradigm of logistics. Nevertheless, since this is moving millions more tons of materials than ever before, one must instead speak of an apparition.

In the world of container transport, this apparition is expressed perhaps nowhere more clearly than on the graphical user interfaces of programs for storage planning like Power-Stow. The representation of each individual container is managed through container icons. The ability to display them in a variety of resolutions (much like computer games)—from the big picture of the ship to the individual container and its cargo—suggests that the movement of the real, weighty steel boxes is just as effortless as the movement of the icons on the screen, especially since the program and the loading events are in fact directly connected through electronic control (or at least monitoring) systems.

Here, the relation of the symbolic sign to the material signified or the addressee seems like an increase in the power of those modern administrative practices that once contributed significantly to the rise of modern bureaucracy: diagrammatic operations in positional systems and the placement and management of material assets in tables and book pages. The mastery of the "large space

Computing with Containers

of transport speeds" is substituted "by the mastery of calculations on paper," as media scholar Bernhard Siegert shows.[10]

Today, the computer-guided, Internet-based control of the container system entangles the large and small spaces anew. On the one hand, the global timing of machine-controlled container transport has allowed the great space of transport speeds to become continually smaller and ever more precisely calculable. On the other hand, the worldwide cross-linkage of all data relevant to transport through the Internet and the configuration of each individual container with a distinct address stretches the small space of the office across the entire globe (at least everywhere there is an Internet connection).

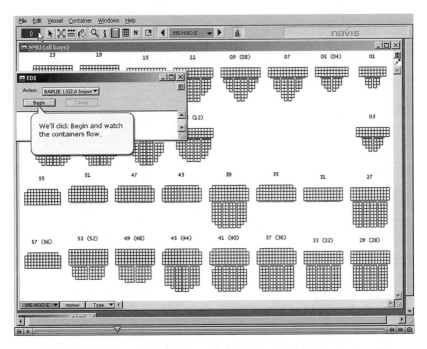

"We'll click Begin and watch the containers flow." Large space and small space united on the graphical user interface: screenshot from an online demonstration of the stowage planning program PowerStow. © Navis.

Light Technology

Back to Project Blackbox. After an advertising tour through 73 cities on four continents, where more than 12,000 potential customers and representatives of professional circles inspected the prototype of the data center container, Sun Microsystems brought the first production model to market about one year later. Now renamed the Sun Modular Datacenter S20 (MD S20), the container held eight standard data racks like those used in classic data centers, in which a total of 280 computer units could be housed. This represented an enormous amount of storage and computing capacity in an incredibly small space, approximately four times more densely packed than usual. This was made possible by a sophisticated water cooling system with access to each individual computer unit, among other things.

In principle, the MD S20 was nothing but a refrigerated container. From the beginning, *reefers*, special containers for the transport of sensitive goods, have represented the technical high end of the container system. In the latter half of the 1960s, computer technology was implemented in them to monitor and control the temperature and air humidity within the container. Essentially, two distinct types of systems have developed.

The are self-contained systems, which have all necessary control electronics built into the container; the MD S20 would be counted among these. They simply need electricity (and in the case of the MD S20, a great deal of water for the cooling system) in order to function. The second are systems that are connected to an external computer monitoring system. Here one must differentiate between reefer containers with *controlled atmosphere*—which produce a completely artificial atmosphere and which "put products in a kind of deep sleep," as it says in a

"Refrigerated Cargo Special" in the *Hamburg Port* magazine—and those with *modified atmosphere*, which intervene based on external conditions in order to maintain an optimal climate for the particular cargo. On most container ships and terminals, there are special places for the refrigerated containers where they can be connected to the electric and/or computer system.

Without temperature regulation, the computers within the MD S20 would overheat and burn out in no time. According to manufacturer specifications, the container can be used with external temperatures of between –22 and 130 degrees Fahrenheit and with up to 99 percent humidity. One of the eight racks is reserved for the cooling system and maintenance computers, which monitor conditions in the container and constantly maintain an appropriate climate for the performance and durability of the computers as well as for the network distributors that coordinate the internal network and the connection of computer performance to the outside.

The design of the Project Blackbox container, introduced in the initial advertisement campaign in matte black with a green company logo, was presumably only for the purposes of recognizability and correspondence between name and appearance. The MD S20s installed to date—on the campus of a California university, on the grounds of a university clinic in Holland, at a mobile phone company in Moscow, and in a factory in India, among others—are maintained in a much more climate-appropriate white, like the models on display on the company's website.

The future prospects of the MD S20 containers are somewhat unclear, however, since Oracle's acquisition of Sun Microsystems in 2010. Oracle officials say they will keep on building the data containers, but they are not promoting them anymore. However, in the process of thinking about the future of data centers, other firms also came to the idea of the container. For instance, there is the ICE cube container from Rackable (ICE is the abbreviation for "integrated concentro environment"), the FOREST container from Verari (FOREST stands for "flexible, open, reliable, energy efficient, scalable, and transportable"), the POD container (portable optimized container) from Hewlett-Packard, and the PMDC (portable modular data center) from IBM.

All were brought to market between 2007 and 2009, and all are based on 20-foot and/or 40-foot ISO containers. Microsoft had the entire first floor of an enormous new data center near Chicago outfitted with more than 100 40-foot data containers. According to the company's figures, each of the containers holds up to 2,500 server computers. As noted in press releases, this arrangement is said to permit 10 times more data density than conventional data center architectures while significantly increasing energy efficiency.

In 2010, Microsoft even won a Green Enterprise IT Award for this facility. The fact that energy efficiency, or power usage effectiveness, and "eco-computing" are the primary focus of discussion in connection with this new generation of network nodes points to another aspect that is perhaps more revolutionary than all the technical innovations: it is apparent that computers, which were seen as the epitome of clean technology for decades, use energy and resources. Indeed, they do so to such an extent that those who until recently had trusted in the insignificance of this aspect compared to the earlier technologies of high industrialization would have to rub their eyes in disbelief.

According to experts, today there are now more than 5 million data centers worldwide. Now information and communications technologies are among the world's largest users of energy.

Computing with Containers

A look inside an MD S20 data center container. © Sun Microsystems, https://photos.sun.com. Used with permission from Oracle.

In Germany, 10 percent of energy use already goes to their calculations, according to a 2009 brochure from the Federal Environmental Agency. Globally, carbon dioxide emissions from their production and use are approaching the level of air traffic.

Since the beginning of the global networking of all computers through the Internet and particularly since the new economy boom of the late 1990s, millions of miles of cables have already been laid. Media theorist Steven Graham demonstrates, with the example of New York, that a view into the underground of a large city shows that the aboveground paths multiply in the various historical layers of the infrastructural systems buried beneath the asphalt. Regarding the catchphrases of the information age or the network society that suggest broad immateriality, Graham writes the following:

> We see that the "information age," or the "network society," is not some immaterial or anti-geographical stampede online. Rather, it encompasses a complex and multifaceted range of restructuring processes that become highly materialized in real places, as efforts are made to equip buildings, institutions, and urban spaces with the kinds of premium electronic and physical connectivity necessary to allow them to assert nodal status within the dynamic flows, and changing divisions of labour, of digital capitalism.[11]

Regarding quantities as well, the networks of supposedly massless and purely virtual computer technologies and the material side of cyberspace are thoroughly comparable to monumental aboveground technical infrastructures like road or rail networks. Indeed, an individual yard of cable is much lighter, thinner, and less material-intensive than rails or asphalt. However, given the unbelievable volume of cable laid, the distinction dwindles. The individual automobile surely uses more energy and produces more pollutants than a computer. But it does not spend the entire day on standby, eating up electricity and producing heat like computers in a data center, in constant use, day and night.

In this respect, the light metal networks of fiber optic cables and silicon compounds are entirely comparable to the heavy metal networks of (container) transport. Particularly as there is a notable narrative strategy for both that has been applied for

Computing with Containers

Widespread talk about the seamlessness of the container system evokes an image of immateriality, like that attributed to information technology. The low price and highly optimized efficiency of container transport suggest the fallacy that material expenditures are significantly lower than before. The opposite is the case. It was only the tremendous multiplication of transport capacities that allowed costs for individual pieces to become so low that the length of the path traveled no longer played a significant role compared to other factors. The fact that the model computer and telecommunications technologies are by no means immaterial matter always becomes apparent where infrastructures spring up, as here in the case of a repeater station in a container in the Sudanese desert. The Internet data line's signals must be boosted every 60 miles or so. Although each carrier (each supplier of data transfer capacities) is responsible for the signals that it manages, in the case of long data lines that are used by multiple carriers, there are often entire container colonies of such repeater stations.

many decades in dealing with the newest technology—a story of becoming ever lighter, more fleeting, more immaterial, the disappearance of tactile references. The modern myth of immaterial technology fantasizes about immateriality and eternal, almost lossless readiness, where it is actually an abstraction. The principle of the black box is a fundamental element of this mythology.

Today, we have to deal with technical systems everywhere in order to control them, whether it be a car, a computer, electricity or a power outlet. And the functions of these systems we neither know nor have to understand (and often could not understand, even if we wanted to). Philosopher Hans Blumenberg, who dealt very essentially with developments in modern technology in the second half of the twentieth century, writes about this situation of fundamental and constitutive misjudgment:

For the sake of this suggestion of ever readiness, the technical world is a sphere of housings, paneling, nonspecific facades and camouflage, independent of all functional requirements. The human share in function is homogenized and reduced to the ideal minimum of pressing a button.

In fact, dealing with a great many technologies has become easier in the last century or so, enhanced since the introduction of computer control systems. However, this finding is true only for the front end of technologies, for the intersections and user interfaces, whether push buttons, control panels, or computer screens. On the back end, behind the fixtures, there is still heavy equipment, technical infrastructure, machines, and machine combinations, whether archaic power plants or shipping transport lines, gas pipelines, automatic hydraulic hoists and grapplers for containers, or hypermodern data centers whose individual components may be comparatively small and light but whose tremendous quantity puts them into the dimensions of factories.

Computing with Containers

At the interface between physical and symbolic reality: container bridge simulator, port of Bremerhaven. Film still from Olaf Sobcak, dir., *A.G.V.– T.E.U.*, documentary film, 2007. © Olaf Sobcak, Maik Freudenberg.

Self-Controlling Logistics

Creating a new record on an object means doubling or multiplying its existence. In principle, this is true for every historical system of inventory and accounting, perhaps beginning with the ancient Egyptians. With the dynamic storage of such records on the Internet and on all manner of temporary (and mobile) carriers, the data no longer rest securely in a computer but rather are distributed and duplicated according to principles like so-called ocean storage on a network of global computing networks and various storage media; this principle seems to be magnified. Things acquire a form of subjectivity and a capacity to act such that their records have a life of their own.

Proof of this new development are visions like those of self-controlling logistics and an Internet of things (in which every good would receive its own address and identity on the Internet). This is the culmination of the notion of light technology, of matter that organizes itself according to our needs, which have been programmed into it. It is Goethe's *Sorcerer's Apprentice* reloaded.

On the one hand, both visions are closely tied to the seemingly unlimited expansion of digital storage capacity that makes it theoretically possible to assign each of the billions and billions of objects coursing through the world its own Internet address and therefore its own unique identity. On the other hand, they are directly connected to the massive production and extensive dissemination of very small storage and transfer media like radio frequency identification (RFID) chips, which can transmit data without cables over a certain distance while using comparatively little energy.

In 2004, two research centers were founded in Germany for the development and testing of intelligent logistics systems: the LogMotionLab of the Fraunhofer Institute for Factory Operation and Automation in Magdeburg, and a collaborative research center financed by the German Research Foundation at the University of Bremen. At the latter, production technicians, economists, computer scientists and electrical engineers probe the possibilities of a decentralized self-controlling logistics network. Analogous to the transport of data packets on the Internet that find their own way according to existing capacities, material packets should also be put in a position to be able to organize the precise transport process between two given places independently.

To this end, each loading unit and each means of transport receives an agent in the Internet with which they communicate via RFID and the cellular network system. For their part, the agents can make contact and negotiate prices and routes. In this way, according to the vision, transport processes could come about whose complexity exceeds the capabilities of human coordinators as well as centralized computer systems. This is the current variant of what remains of the once lofty notions of cybernetics—intelligence no longer arises through the construction of a centrally controlled omnipotent computer unit but now arises through the networking of microintelligences with very limited decision capabilities.

These collectively form a system of higher complexity, after the model of an ant colony swarm. However, if the objects dissolved in such a way into a network of microsubjectivities, there would once again be an apparent fundamental reversal, this time from the tangibility of the transported item as a material object to its constitution as a framework of information about itself and relations to others. "The things themselves become logistical processes," media scholar and logistics researcher Christoph Neubert writes.[12]

The culmination of this scenario is a neighborhood-level system of local area networks with intelligent containers. Each container is outfitted with a minicomputer and an RFID reader, which makes it possible for the intelligent packets inside to merge into a local area network and which produces a connection to its adjacent containers. In this way, a local network of containers takes shape in which each functions as a server for a local network of packets inside. The topmost container on a stack transmits and receives information to and from the Internet via satellite at regular intervals. Thus, on the one hand, the complete tracking of each individual packet around the globe would be possible. On the other hand, the individual loading units could spell out their continued transport in a timely manner before their arrival.

Even if containers could reposition themselves again as meta-communication units, in a certain sense, the scenario of self-controlling logistics would signify an exit from the container system. For if the mastery of even the largest volumes of data—and the objects that generate them—no longer poses a serious technical and organizational problem, a main reason for consolidation and enclosure has disappeared. Transport could pick up where it left off before containerization, with the individual cargo, and one could consider alternative transport units, such as industrial loading according to the kind of bulk commodity.

In the meantime, a damper was put on the lofty plans because RFID technologies have spread with much less speed than predicted a few years ago. Nevertheless, in certain areas of the logistical system, self-control is already a reality, as in the automated high-bay warehouses and distribution centers of the large retailers already discussed in chapter 5. Here the balance of power between humans and objects has been completely reversed: intermediate human agency is no longer necessary to give instructions. The parcels themselves—facilitated by RFID chips and computer communications systems—command how they are to be handled and where they are to be brought.

Container Formats

"Containers are everywhere; they are the true globe-trotters, the Lego blocks of globalization, the zip file of world commerce. Packing, shipping, unpacking, compact and speedy." This is how author Ralf Hoppe describes them in a November 2001 article for the German magazine *Kultur Spiegel*. However, the analogy also works in the opposite direction. If the containers are said to lose meaning as a concept on the level of physical transport, for the time being they maintain a central position in precisely that place that is primarily responsible for their loss of meaning.

By a wide margin, most data that are processed through the global high-speed lines of the Internet on a daily basis are audio and especially video files that are saved and transferred by so-called container formats. As a study on worldwide Internet use performed annually since 2006 shows, more than two-thirds of total data traffic goes to the computing of P2P file sharing, direct download, and audio and video streaming connections, through which container format data are exchanged: films, music, pornography, programs, games, and so on. Thus, one can say that containers have taken the lead not only in the global transport of goods on ships, trains, and trucks but also in data through the Internet. To carry the analogy further, the amount of Internet container transit measured in bits even exceeds the share of containerized goods in global transport measured in tons.

Among the most well-known container formats are AVI, MPEG, MOV, VOB, MKV, AIFF, WAVE, DivX, SWF, FLV, and PDF. Typical of these is the fact that they integrate different types of data; they only define the nature and structure of how the contents are to be stored or transported and distributed rather than accessing these contents themselves. Therefore, a container format for the storage and playback of video data may regulate the synchronization of video and audio data, but it can use a whole spectrum of different data formats and codecs, computer programs that are responsible for the encoding, decoding, and compression of the files and thus for the actual processing of the data. The information necessary for this is found in the header, the first few command lines of the file. Thus, there are multiple program levels operating over one another, while the container format acts as an outer shell for the coordination and corruption-free storage or transport of data packets.

The transport metaphor continues on different levels of hardware and software. For instance, films, pieces of music, and similar data that were saved on a physical, portable (and sellable) medium like a CD or a DVD are called *packaged media*, in computer jargon. And while standardized steel containers are switched from one carrier to another in the physical world, indifferent to their contents, the world of information transport via the Internet and mobile phone networks is based on the switching of packets, on the indifferent distribution of split data packets, whose destination and precise composition are defined by a few lines of standard protocol.

These are not called containers, but the structural similarities with physical transport are nevertheless striking, especially as the identification and distribution of the steel boxes—with the implementation of RFID technology for container recognition by way of radio waves—is increasingly controlled by a few lines of electronically transmitted code.

Cloud Computing

Project Blackbox is part of a development that computer experts regard as the dawn of a third age in the history of computers. In the first phase only a few companies and institutions operated the enormous and tremendously maintenance-intensive computers, and in the second phase the technology spread to every office in the early 1980s with the introduction of the personal

Computing with Containers

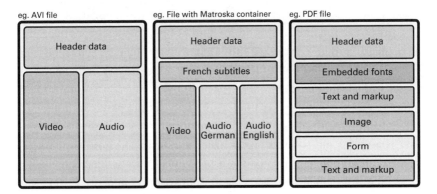

A few examples of container formats. From left to right: AVI, data with a Matryoshka container, PDF. © 2006 Ernst Michael Rohlicek. Also available under Creative Commons BY-SA at http://wikimedia.org.

computer. The coordination of the computers in the internal network as well as the larger storage and computing capacities was provided by data centers that every company arranged according to its own standards.

In the third phase of computer history now being advertised—associated with concepts like software-as-a-service, virtualization, and cloud computing—computing services, storage capacities, and programs are being outsourced to third-party (or network) providers. Rather than each company maintaining all the necessary capacities with its own computer infrastructure in specially constructed or rented spaces—and having to constantly renew and expand in keeping with rising demands—it draws on the resources of cloud computing providers on demand.

These, in turn, must be able to react quickly and flexibly to changing demands. In light of this, the traditional model of stationary, firmly installed data centers is no longer sufficient under certain circumstances and must be supplemented by new, expandable models. It is here that the providers of mobile, modular data centers see their opportunity, since these can be installed on the roof of a building or outside the city in a green field without much effort.

Cloud computing is part of the advance of third parties into the realm of computer networks and computing power. If it was the hardware and software companies like IBM and Microsoft—that is, the producers of machines and of uses for these machines—that dominated the market and defined developments in the first and second phases of computer history, the current leaders are Internet companies like Amazon, Google, or Facebook. They do their business with Internet-based consumer and communications offerings, with meta-applications like search engines or communications networks that span existing infrastructures.

The prerequisites for such an offer are technologies of virtualization. Each of the many users of small sections of large computers is granted the appropriate capacity, and in the process the illusion is created—through graphical user interfaces and user profiles—that he or she alone has access to the resource. Sun Microsystems makes its push into the containerization of data centers explicit in this context. The slogan for Project Blackbox is "the world's first virtualized data center."

Nicholas Carr, author of a much-discussed 2008 book entitled *The Big Switch: Rewiring the World, from Edison to Google*, compares these new, modularized data centers to a prefabricated electric generator system that the great inventor and pioneer of electrification, Thomas Edison, brought to market at the end of the nineteenth century. Under the heading "Trailer Park Computing," Carr comments on his blog (www.roughtype.com) as follows:

Computing with Containers

In many ways, the containerized data center resembles the standardized electricity-generation system that Thomas Edison sold to factories at the end of the 19th century and the beginning of the 20th. Manufacturers bought a lot of those systems to replace their complex, custom-built hydraulic or steam systems for generating mechanical power. Edison's off-the-shelf power plant turned out to be a transitional product—though a very lucrative one. Once the distribution network—the electric grid—had matured, factories abandoned their private generating stations altogether, choosing to get their power for a monthly fee from utilities, the ultimate black boxes.

In February 2007, Google patented a concept through three of its employees that allows ISO containers filled with computers to find their way home in a certain sense, even if *home* means something entirely different here—namely, permanent traveling, placeless at sea. Under patent number US 2008/0209234 A1, the guidelines and technical framework were provided for a water-based data center that provides itself with electricity. The computer facilities are housed in standard containers, the electricity is gained from the sea—the tides, waves, and ocean currents—or from other renewable energy resources, and so is the water for cooling the computers.

Here, as with all the other modular data centers, the underlying idea is that one can make computing capacity available where it is needed quickly and flexibly. As a rule, the transport could be managed easily with the existing container logistics structure—with the decisive difference that the container is installed not on land but rather on board a ship or a data raft located three to seven miles offshore. The Internet connection is either produced by hooking up to one of the undersea cables or through strong wireless network connections to receivers on the mainland that then feed the signal into the Internet. A satellite connection would also be possible.

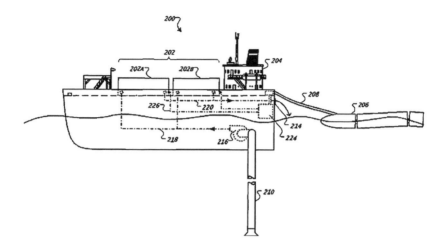

Patent drawing of the data raft designed by Google staff. © Jimmy Clidaras (Los Altos, CA), David W. Stiver (Santa Clara, CA), William Hamburgen (Palo Alto, CA), Google Inc. (Mountain View, CA), Water-based Data Center, U.S. Patent 20080209234, filed February 26, 2007, and issued August 28, 2008.

Thus, the data center containers would be at sea, in the space that provides more than any other the metaphors for the medium that runs through its circuits: the Internet. Such offshore data centers would raise interesting legal questions about the proper jurisdiction of the data saved on them. On the one hand, the situation is reminiscent of the diverse offshore locations in the financial world where companies have their pro forma headquarters in order to save on taxes. Even so, in those cases, these are islands or continental microstates. On the other hand, the data raft project makes one think of the ongoing, vigorous debate over Internet piracy, the systematic violation of copyright terms through the exchange of software, music, and videos made possible by peer-to-peer network providers like the Napster and Bit-Torrent search engines of several years ago and the Pirate Bay more recently.

In January 2007, the operators of the Pirate Bay (who were subject in early 2009 to serious penalties from Swedish authorities in a trial followed with great interest internationally) attempted to purchase the self-declared micronation of Sealand, situated on a former English World War II defense platform in the North Sea, six miles off the coast of Essex, in order to move there with their servers. However, the deal faltered amid the resistance of the "sovereign" of Sealand, a British army major who himself had occupied the platform in 1967 because he had wanted to operate a pirate radio station from there.

In its compressed worldliness, the planned data raft also conjures up associations with the raft in Neal Stephenson's novel *Snow Crash*. In the 1992 book, Stephenson precisely describes a whole array of possible developments in the computer age—above all, a three-dimensional version of the Internet, the "metaverse," and all conceivable forms of electronic, radio, and laser-operated monitoring and recognition technologies. He designs a dystopian vision of an America fallen into nothing more than privately managed and guarded "franchise-organized quasi-national units."[13]

There is nothing to remind one that there could have once been a connected society. All around there are only hollowed out forms: former government institutions as brands, radical cultural and racial segregation in the guise of consumerism, and microterritories between which greed and violence rule. It is a world of warehouses, containers and corridors, gated communities, and privatized transport and communications networks. As a parallel world of experience, there is the metaverse, at least for those who can afford to access it, but where things are not principally different.

Vis-à-vis this continent decayed from any uniform order into an island archipelago, Stephenson's dark vision sees "the Raft" bobbing up and down. It illustrates a grotesque compression and increase of segregated diversity on land, an agglomeration of various ships, houseboats, rafts, and hijacked yachts constantly changing in number and form. Inside, there are a few container ships, with the core consisting of an oil tanker and the aircraft carrier *Enterprise*. Together, they form a floating city, populated by refugees, pirates, and dubious wheeler-dealers from around the world:

> The worst thing that can happen on the Raft is for your neighborhood to get cut loose. That's why the Raft is such a tangled mess. Each neighborhood is afraid that the neighboring 'hoods are going to gang up on them, cut them loose, leave them to starve in the middle of the Pacific. So they are constantly finding new ways to tie themselves into each other, running cables over, under, and around their neighbors, tying into more far-flung 'hoods, or preferably into one of the Core ships.[14]

The raft circles the world with the currents, takes on refugees on one side, only to lose them again in wealthier areas on the other side. The majority of its inhabitants are infected with a metavirus that allows people to speak in tongues, a preconscious form of communication, a universal language before the Babylonian proliferation, that makes them centrally controllable. In parallel, this virus is also spread across the metaverse. Hackers, computer programmers for whom thinking in binary code has become a second language, are especially susceptible to the infection, transmitted through Bitmap files with white noise.

Behind everything stands a media mogul who has allied himself with a cult leader in order to gain control of the entire world. On the raft, the vectors of a world history of liquefaction and control converge—computer and container in network on a ship, at once in an ocean of data and of water. Post-territorial, transnational structures are controlled by protocol in a universal binary code.

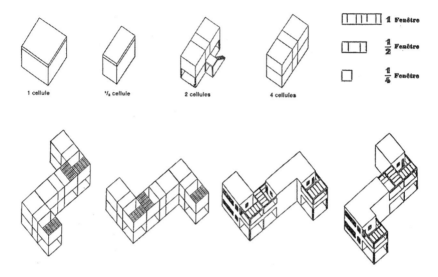

The dream of the modern: rebuilding all of society from standardized room modules. 1923 drawing by Le Corbusier. © F.L.C. / ADAGP, Paris / Artists Rights Society (ARS), New York 2014

7 Life in Cells

How are we to imagine an existence oriented solely toward the Boulevard Bonne-Nouvelle, in rooms designed by Le Corbusier and Oud?
—Walter Benjamin

In 1923, Charles-Edouard Jeanneret, better known as Le Corbusier, published a series of drawings of a house constructed from cells and the combined possibilities that this would present, in *La Maison Standardisée* (The Standardized House). This French architect of Welsh-Swiss heritage had the architectural cell in mind as the smallest living unit for people, from the cell of a monk to prison cells, but also the modern ship's cabin. Furthermore, he consciously aimed at the biological connotations of the cell as the basic building block of life.

Since his training period in Berlin with Peter Behrens, an industrial architect and the preparer of the modern, and with the Parisian architects Auguste and Gustave Perret, who were then introducing the revolutionary method of concrete frame construction, Le Corbusier was convinced that the future of building lay in industrialization, in the use of rationalized working methods and mechanized mass production in the style of the Ford Motor Company's factories. A new architectural aesthetic in keeping with the spirit of the modern times had to be aligned with the great achievements of the engineers, with concrete silos, steel bridges and cranes, ocean liners, airplanes, and automobiles.

In this way, he came to the prefabricated, standardized modular room. As a concept, this was nothing more than a container to be lived in, *avant la letter* [before the term was invented]. By showing cells over empty space at one time and over filled space at another, Le Corbusier demonstrated a central element of the container stacking principle that was brought to bear in the cell structure of container ships or in the construction of a container terminal, long before its technological realization in transport logistics. The space and position of the containers are always defined and present, whether they are real or only virtual. The container principle allocates standardized units of space and thus apportions them according to economic calculation. Whether they are filled with air or with valuable cargo—indeed, whether they themselves are present or absent—is secondary. It is primarily about a form of spatial processing.

This programmatic shift in modern architecture toward logistical production processes on the model of the American automotive industry, on the one hand, and toward models and metaphors of spatial organization from the realm of modern transit and transport, on the other, awakens the following suspicion: Could it be that since then, houses are no longer lived in but rather occupied or loaded much like vehicles? Thus, the controversy regarding the cell and the stacked house constructed from cells or living containers, Le Corbusier's "living machine," became a pivotal point in the political and aesthetic disputes over modern industrial society. Since its "heroic" phase after World War I, the battle over and with modern architecture has been a fight with the idea of living in cells—constructive, programmatic, and ideological. A comprehensive treatment of this subject would fill a book on its own. Consequently, this chapter limits itself to highlighting the container shapes of modern architecture and its associated programs.

The Death of the Modern

The notion of tabula rasa (blank slate), of being able to put an end to the loathsome past and make a radical new beginning, is one of the starting points of modern architecture. Its agents hoped to create the conditions for a better society by bringing about a crystalline, rational order that optimally organized basic human needs. In this way, they came upon the simplest spatial forms: the cell as a serializable unit of space and the grid as a universally applicable pattern for organization. Like the other grand Tayloristic visions of rationalizing all kinds of human environments and activities (see also chapter 5), their programs were formulated between the two world wars and implemented extensively in the postwar period. In the 1960s they found themselves in crisis.

The architectural critic Charles Jencks, one of the theoretical founders of the postmodern, described the beginning of the end of a high-rise project that had been constructed only 20 years before with the highest hopes of urban planning:

> Modern architecture died in St. Louis, Missouri on July 15, 1972, at 3:32 p.m. (or thereabouts) when the infamous Pruitt-Igoe project, or rather several of its high-rise buildings, were given the final *coup de grâce* by dynamite. Previously it had been vandalized, mutilated, and defaced by its black inhabitants, and although millions of dollars were pumped back, trying to keep it alive ... it was finally put out of its misery.[1]

Apart from the latent racist undertones of his report, which signal the failure of the universal claims of modernism on another level, in its exaggerated formulation of the death of the modern, the critique is primarily directed at modernism's belief in rationalization—and against the principle of an elitist building culture, dictated from above and unresponsive to the actual needs of residents and local conditions. Against an architecture and style of urban planning that sees itself as a revolutionary surprise attack and that goes hand in hand with a technocratic planning process present in all the world's modern and modernizing societies in the years after World War II.

Nevertheless, what is interesting (and characteristic of the entire approach of the postmodern) is the fact that Jencks ultimately reduced his fundamental critique to aesthetic aspects. Lewis Mumford, an American historian of the city and technology, denounced modern architecture as cardboard boxes, shoe boxes, egg cartons, and file cabinets and lamented the dystopian spirit of George Orwell's *1984* with these metaphors of the mechanistic and of standardization. From critiques like this, Jencks drew the conclusion that architecture must choose other metaphors in its formal language. In order to function, it must fulfill its communicative task; it must create symbols with which its residents can identify.

Largely untouched in such an approach are the social, legal, and political conditions under which architecture arises and persists and shapes the lives of its inhabitants as a built environment. On the one hand, this huge blind spot in postmodern reform speaks eloquently to the exaggerated self-perception of the generations of architects and designers who followed the heroic moderns, who persuade themselves that it is only how something is shaped that determines whether it is good or bad for society. On the other hand, this approach facilitated the solidification of a truism that is widespread to this day, even though it neglects the structural and decisive functional aspects—namely, the comfortable belief that the container and box architecture of high-rise ghettos is largely responsible for the social ills within them.

Blow It ... Up!

However, regardless of how they are built, whether they are stacked serially from prefabricated cells or built according to conventional methods with brick and mortar, it seems much more relevant how they are managed. Thus, it was one of the most astonishing findings of German reunification that many of the East German slab housing complexes, which had been assumed to be the embodiment of inhuman technocracy, were well accepted by their inhabitants and allowed a high degree of social mixing—at least as long as the more well-to-do residents could not simply move away into nicer, newly renovated quarters of the old city or areas with single-family homes.

In contrast, in the large high-rise buildings on the fringes of Paris, where regularly, to this day, small revolts break out with burning cars and looted shops, the residents are as isolated spatially as they are socially because of their lack of transit connections to the inner-city areas. (Much the same goes for many of the housing projects in U.S. cities, which are infamous for being dangerous because of the high amount of mostly drug-related crime.) The promise made upon construction of these complexes, that they would be connected to the local transit infrastructure, was never kept. Like a modern-day city wall, a six-lane highway ring lies between them and Paris.

The same principle of administrative neglect was at work in the aforementioned Pruitt-Igoe complex. Progress or decay—St. Louis must choose; this is how the large-scale renovation project was presented to the public in 1950. For the construction of the large structure, a run-down inner-city quarter was completely razed, and in its place, 33 identical living sculptures were erected with a total of 2,780 residential units. The buildings were designed by the architectural firm of Hellmuth, Yamasaki and Leinweber on the model of Le Corbusier's living machine, with courtyards on every third floor from which all apartments were accessed by way of (unlit) staircases and landings on the interior of the building.

In 1951, the design won a prize from the American Institute of Architects, which became one of the largest architectural firms in the United States. It wanted to create a new model of public space called *horizontal neighborhoods*, a vertical organization of residential cells into a three-dimensional grid modeled after the modern city. Instead, the firm achieved a vertical organization of horizontal slums.

Public drug dealing and prostitution on the landings, and vandalism and criminality rather than community cooperation, led to a situation in which no one who could afford to move elsewhere wanted to live there. Meanwhile, the city of St. Louis lost all interest in the social living project that had begun with so much ideological investment. For instance, the city refused to pay for trash removal in the complex simply by declaring its pathways to be private streets. In the early 1960s, instead of the improvement of infrastructure and decent management being the focus, social workers were sent over. When their involvement, then studies conducted by the University of St. Louis on improvements in the living environment, and then a more than

$7 million U.S. government renovation program did not lead to improvements, the residents were consulted for the first time. "Blow it ... up! Blow it ... up! Blow it ... up!" was their radical vote, as author Tom Wolfe notes in his book *From Bauhaus to Our House*, his wonderfully written history of modern architecture in the United States.

Since then, the image of new buildings collapsing has stood as a symbol for the failure of the modern as well as for its contradictory dynamic, which makes destruction into a constructive principle and whose high-flying visions regularly end in a dark abyss.

The Destruction of the Box

Even before architectural modernity had really begun in the capitals of Europe, and with it the apotheosis of basic geometric elements, the demand for their demolition existed. In the beginning of the twentieth century, Frank Lloyd Wright, the creator of the prairie style, much admired even in Europe, preached from the wilds of the North American prairie about a method of construction based on what he called the destruction of the box: the dissolution of the Western building tradition of rectangular forms reaching back into antiquity, made possible by new building materials (namely, reinforced concrete), leading to open floor plans. In this sense, Wright's vision was very comparable to one of the fundamental concepts of Le Corbusier, the free floor plan.

In a lecture that Wright gave at the American Institute of Architects in New York in 1952, he explained in retrospect how he came to what may be his most important concept. Initially, he framed the "war against the box" as a political task:

Down all the avenues of time, architecture was an enclosure by nature, and the simplest form of enclosure was the box. The box was ornamented, they put columns in front of it, pilasters and cornices on it, but they always considered an enclosure in terms of the box. Now when Democracy became an establishment, as it is in America, that box-idea began to be irksome. As a young architect, I began to feel annoyed, held back, imposed upon by this sense of enclosure which you went into and there you were—boxed, crated. I tried to find out what was happening to me: I was the free son of a free people and I wanted to be free. I had to find out what was the cause of this imprisonment. So I began to investigate.[2]

He then proceeded to explain the means that he found to eliminate the "boxyness" of architecture, with the first deliberate example being the 1904 construction of the Larkin building (the administrative building of a soap manufacturer in Buffalo, New York). He found "a natural path to liberation" in setting the four staircase towers apart from the main building and making them into their own architectural elements. In so doing, the floor plans were free for a stand-alone design independent of conventional constructive necessities (derived from statics):

Instead of post-and-beam construction, the usual box building, you now have a new sense of building construction by way of the cantilever and continuity. Both are new structural elements as they now enter into architecture.... [In] this simple change of thought lies the essential of the architectural change from box to free plan and the new reality that is space instead of matter.[3]

Similar voices were raised in Europe as well— for instance, the German architect Bruno Taut, a figure as colorful as Wright. He became famous through his large housing projects in Berlin, which are seen as exemplary to this day, as well as for his revolutionary, utopian concept of Alpine architecture and the "dissolution of the cities" of 1918–1919, in which he promoted a melding of city and nature. His campaign slogan of "Let them collapse—the constructed vulgarities!" was directed against the nested mass settlements of the late nineteenth-century industrial city.[4]

It is interesting that Wright's obsessive preoccupation with the box (and possibly the obsessive preoccupation of other important architects of modernity as well) can be traced back to a pedagogical system of early childhood education dating from the first half of the nineteenth century, consisting of basic two- and three-dimensional forms, the Froebel Gifts. At the 1876 World's Fair in Philadelphia, Wright's mother came across the educational toys of the German reformist pedagogue and kindergarten pioneer Friedrich Wilhelm August Fröbel (1782–1852), and she bought her then five-year-old son a set. Wright explained this connection near the very beginning of his career, noting that the coordinating effect of the grid in combination with standardized components was "completely natural and unavoidable" for him as a design principle: "it is based on the straight line technique of the T square and the triangle. It was inherent in the Froebel system of kindergarten training given to me by my mother."[5]

The dialectical movement between the formative and disciplinary power of the basic geometrical form of the box and the various attempts at its conquest can be seen as characteristic for the entire history of modern architecture, from its beginning around 1900 through its many deaths and resurrections in the last third of the twentieth century to today, from deconstruction to amorphous blob to network architectures.

The housing shortage and the spatial limitation led to the grand, ambitious housing programs, in the course of which Taut's complexes and other successful examples of social building were constructed in the 1920s (and which have never played a comparable role in the United States; Wright developed his

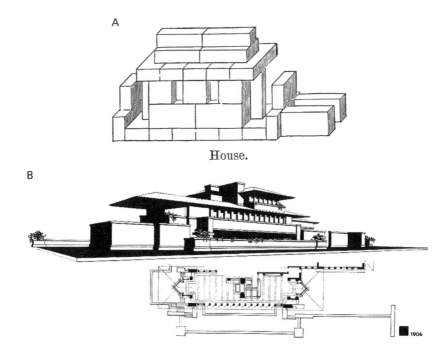

(a) Froebel Gifts, 1877, and (b) plan for the Robie House by Frank Lloyd Wright, 1906. The combination shows the remarkable path of a pedagogical toy, from building blocks to avant-garde buildings. © 1996 The Frank Lloyd Wright Foundation.

prairie style in Chicago, a city that could spread out almost without limit over hundreds of miles). These are the fundamental reasons that the box returns as a conceptual element after World War I in the form of a cell, to become universally generalized after World War II.

Honeycombs

In 1922, Walter Gropius, the founding director of the Bauhaus school of design and the most important early advocate, other than Le Corbusier, for the industrialization of building, made designs for a serial house together with Adolf Meyer and Fred Forbat, for the Bauhaus residential development Am Horn in Weimar, Germany. Much like a model kit, the various types of houses based on standardized individual spatial units could be put together in any combination, according to the needs of the inhabitants.

One year later, these drafts were shown in an exhibit along with models and drawings. A few years later, Gropius published one of the sketches again, in an article entitled "Standardization and Housing Shortage," which was the print version of a lecture that he had given a year earlier at the annual meeting of the German Standards Committee (see figure on next page). A few introductory sentences on the subjects of building normalization and the savings potential of this "fundamental groundwork of rationalization" culminated in the following declaration:

With a wise restriction to a few types for the buildings and things of daily use, their quality increases and their price decreases, and thereby the entire social level is necessarily raised. If we succeed in providing the majority of people with cheaper and better housing through the application of these principles, then one of the most important basic economic questions would be resolved.[6]

In this way, he formulated the entire program of social engineering that combines with industrial building, normalization, and rationalization. In the text for another image from the 1923 exhibition—which contains a drawing of the building site, two layout drawings of the planned structures, a photo with models,

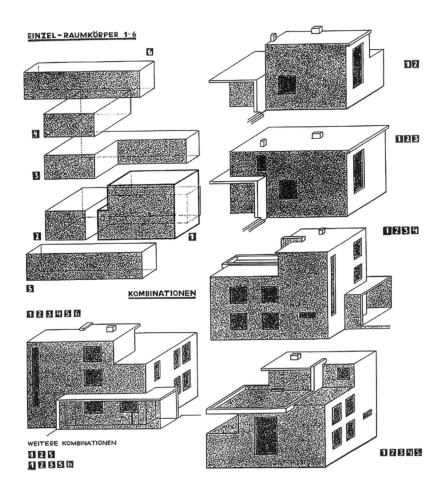

Walter Gropius, Adolf Meyer, and Fred Forbat, 1923. Architectural drawing for the planned Bauhaus residential development Am Horn, Weimar. From Walter Prigge, ed., *Ernst Neufert: Normierte Bau kultur* [*Ernst Neufert: Standardized Building Culture*] (Frankfurt am Main: Campus Verlag, 1923), 330.

and a schematic sketch of the standardized spatial units and their possible combinations—he shows that he was inspired by a source other than Le Corbusier and the (American) industrial culture. Under the heading "Honeycomb Construction," the exhibit states, "High variability of the same basic type through the systematic addition of attached room modules." The houses in the community in Weimar were never built, but the plans served as a basis for the modular homes built from prefabricated concrete slabs constructed a few years later in the experimental settlement of Dessau-Törten.

The bee metaphor as an ideal descriptive schema for the construction of a society was already well established at this time. If in the debate over bees as a model for state organization around 1900, organic perspectives on the relationship of the individual to the greater whole were given precedence (a view that regained great popularity under Nazism), the view shifted under the influence of industrialization to the aspect of architecture and the serial production of living space.

Thus, the visionary description of the modern city that Walther Rathenau, a German industrialist, politician, and universal commentator, gave in 1912 culminates in the "stony image" of "honeycombs" cast in concrete, from which the new cities were being constructed—with worldwide uniformity—whether "workplaces, ... homes or ... resting places" of the "international world camp":

All larger cities of the Western world are identical in their structure and mechanics. Resting in the center of a spider web of railways, they shoot out their petrifying road fibers all over the country. Visible and invisible networks of moving traffic pull and plough through the street ravines and pump human bodies from the extremities to the heart twice a day. Second, third, and fourth networks distribute water, heat and power; an electrical nerve beam carries the vibrations on the mind. Nutrients and irritants glide on rails and water, used material pours out through channels. When viewed as a cross-section, the stony picture is the same everywhere: honeycombs equipped with subtle substances, paper, wood, leather, fabrics, arrange themselves in sequential order, outwardly supported by iron, stone, glass, concrete.[7]

Gropius was therefore not alone in his analogy of honeycombs, nor was he only speaking metaphorically. When the two most important founding figures of the German Institute for Standardization (DIN), Walter Porstmann and Wilhelm Ostwald, made their calculations that would lead to the introduction of the DIN-A4 standard for paper formats in 1921, they faced a mathematical problem arising from the calculation of rhombic dodecahedrons—that is, the form of honeycombs, the space-saving three-dimensional organization.

Regardless of any aesthetically or politically motivated criticism, the idea for the serial organization of human living and life in honeycombs survives to this day, both structurally and aesthetically. Philosopher Peter Sloterdijk writes the following in the third volume of his vast Spheres trilogy:

The key to the relationship between the cell and cellular organization that is characteristic of modernity lies in serialism. As the development of the cell takes into account the spirit of analysis through reduction to the elemental level, the building of houses on the basis of such elements represents a combinatorics, or better, a form of "organic construction"— with the goal of creating architecturally, urbanistically and economically sustainable ensembles of modules.[8]

However, the driving principle of this organic aspiration is the production of completely artificial living spaces. In this respect, the standardized modular space, the living container, must be seen as a futuristic project, in whatever traditional masking it may appear. Modern architecture and city planning are subject to a cybernetic guiding ideal. Or, as Sloterdijk puts it, the more "the explication of human residence in man-made interiors" advances, "the more the construction of apartments will resemble the installation of space stations."[9]

Cells at Human Scale

"I see the core problem of modern building in industrialization. If we succeed in carrying out this industrialization, then the social, economic, technological and artistic questions will be easily answered."[10] This statement, strangely naive to modern ears in its belief in technology, came from the mouth of architect Mies van der Rohe in 1924.

Walter Prigge, a sociologist of the city active in the Bauhaus Dessau foundation, describes how thinking in minimum room sizes, grids, and industrially prefabricated cells as a fundamental element of modern architectural designs replaced classical concepts of urban planning and was thus ideally suited as a medium of social reformist and utopian visions:

> The series of residences produced according to industrial principles tends to be unlimited, and transforms the historical residential block to a residential cell. With it, the modern building and thus also the historical building typology is finally detached from the context of street and city that is reconfigured by the montage of living units. Rebuilding the world uniformly from the smallest spatial cell: that is the rational architectural utopia in the mid-twentieth century.[11]

Rebuilding the world from the smallest spatial cell was a core element of Le Corbusier's architectural theory since the early 1920s, propagated in many essays and designs. In 1929, in a series of lectures that he gave in Argentina and Brazil, he summarized the application of his cellular concepts from the single-family house to the planning of an entire city with 3 million inhabitants:

A cell at human scale: 15 m². ... The new forms of standardization, of industrialization, of Taylorization will be exploited ... for the home, the office, the workshop, the factory....

These methods of industrialization as a result of standardization naturally lead us to the skyscraper: its form is determined by the stacking of cells on human scale....

Let us multiply the standard elements of the cell.... The house need no longer be measured in meters [yards]— it must be built in kilometers [miles].

These cells must be stackable by the millions.[12]

The inspiration for the size of the "cell at human scale" came to him in his luxury cabin on the ship that brought him from Europe to South America for his lectures: "A man is happy, lives just like at home, sleeps, washes up, writes, reads, receives his friends—all this in a 15 m² room."[13] (The size of a converted 20-foot standard ship container, used frequently today to house refugees or construction workers and even for chic residential studios, is 14.77 square meters.)

In the context of radical models for the rational assignment of space and the standardization of human living environments, one must mention the Gropius student Ernst Neufert, who is also called the "Taylor of construction." Neufert belonged to the first generation of architecture students at Bauhaus. Subsequently, he was the manager at various construction sites for Gropius for several years. In contrast to most representatives of modern architecture, who feared persecution because of their leftist political sensibilities, their insufficient "Aryan" heritage, or both, Neufert remained in Germany during the Third Reich. He became one of the leading figures in Nazi housing construction and attempted to convince the fascist functionaries that the spirit of Nazism and the new spirit of modern architecture had a common denominator.

While still under Gropius, Neufert had begun to grapple intensively with the possibilities of standardized building. He developed an approach that went from the interior of the house, from the residents and their uses, outward. His *Architects' Data*, appearing for the first time in 1936, presented the results of years of study. It showed how a complete system of architectural norms could be produced for all functional areas of architecture—building upon humans as the measure of all things.[14] Every conceivable space and every conceivable activity—from the cradle to the grave, from a staircase to an airplane, from a bed to a bathroom—received its minimum measurements. Starting with layout grids, Neufert rationalized the various types of use and transformed them into serial spatial schemata. Standard industrial measures for concrete areas of application filled the abstract cell structures of modern architecture with functionalist content, all on the same scale.

Architects' Data became the most successful architecture book of all time. It has been translated into nearly all the world's languages and appears in new editions almost yearly in the German version alone. Whenever certain measurements are unclear in the planning phase, Neufert's work is consulted. No other architectural system, not even Le Corbusier with his system of the modular and his no-less universal claim, managed to define with such breadth how large the (invisible) containers would have to be for people's various activities.

During the Third Reich, Neufert continued his research and preached for a standardized construction method based on the totalization of the grid measurements already constitutive to *Architects' Data*. In 1943, his book *Bauordnungslehre* [*Building Regulation*] was published, with a foreword by his patron Albert Speer, the Reich's construction minister. It was based on a grid that not only applies within a building but also unites all buildings at a

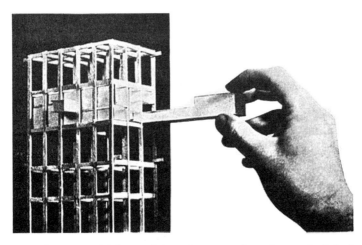

Le Corbusier, model for steel-framed construction of a standard-sized living unit for approximately 1,600 residents in Marseille, 1946. The model for his construction is a wine rack. The individual living units are inserted into the framework like bottles. The hand of the architect that drew the plans and built the models in the imagination is transformed into a large apparatus, a crane that shoves the container apartments into the giant frame of the "machine for living in." From Le Corbusier, *Oeuvre Complète: 1938–1946* [*Complete Works: 1938–1946*], Vol. 4 (Zurich: Les Editions d'Architecture, 1946). © F.L.C. / ADAGP, Paris / Artists Rights Society (ARS), New York 2014.

site and determines their location and external proportions. It could potentially be expanded to global scale and unite everything with everything else. "Just like on the ocean," Neufert wrote in an explanatory caption, it permits one to "immediately and clearly determine the position of the buildings and all other installations. The buildings, laid out according to their dimensions, necessarily fit into this grid."[15] This is the principle of "snap to grid," long before the introduction of Photoshop.

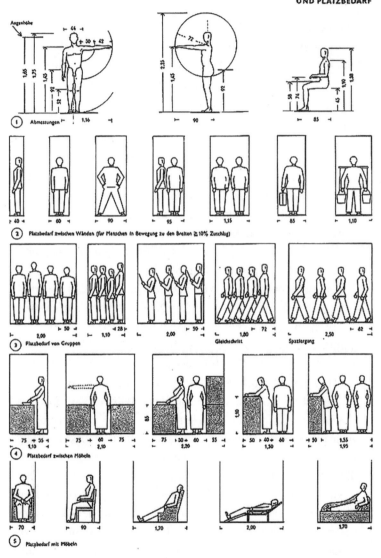

People in functional containers. Cover image of the original edition of the most successful architectural book of all time, Ernst Neufert's *Architects' Data*, 1936. From Walter Prigge, ed., *Ernst Neufert: Normierte Baukultur* [*Ernst Neufert: Standardized Building Culture*] (Frankfurt am Main: Campus Verlag, 1999).

However, Neufert failed in this radical approach to normalization amid the resistance of colleagues, who apparently found his unfettered technocratic nature unsettling. Even in a system that drove the world into a total war and industrially annihilated human lives in camps, people shied away from fitting the entire living environment into a total grid. Yet Neufert's recommendation anticipated what is now a ubiquitous practice, with satellite-aided geographic positioning systems (GPS), on the one hand, and computer-assisted design (CAD) programs on the other. Scaled to the size of Earth, it potentially makes possible the exact localization of buildings. Scaled down to the proportions of the individual buildings and rooms, it defines the size and position of each object within the house, from the walls and windows to the steps and furniture.

Neufert's idea of a unified format must be interpreted as part of a comprehensive process of standardization that began in the first half of the twentieth century and that defines the world today in countless details. This is seen as early as 1912 in the German organization Die Brücke [The bridge]—not to be confused with the similarly named expressionist group of artists. The group, including two of its most important members, Wilhelm Ostwald and Walter Porstmann, who have already been briefly mentioned, made plans for the introduction of a world format for everything associated with writing: books and folders, shelves, and eventually entire libraries, offices, and hotel receptions.

The gridlike layout dominates here as well. The cell-shaped organization comes directly from the folding of the two-dimensional geometric order of the paper into three dimensions—two-dimensional spatial schemata becoming architecture. The plans of Die Brücke do not lead to a world format for bookcases, filing cabinets, and administrative buildings, but they do at least result

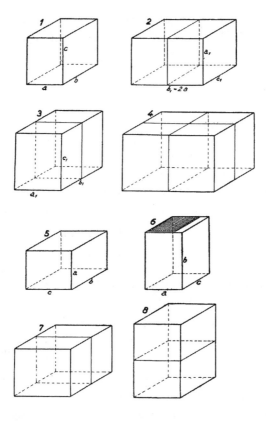

Expansion of the two-dimensional paper norm to the world through folding. Design for DIN space formats, 1918. From A. Heilandt, "Die Vereinheitlichung der Formate auf wissenschaftlicher Grundlage" ("The Standardization of Formats on a Scientific Basis"), in *Mitteilungen des Normenausschusses der Deutschen Industrie* (*Proceedings of the Standardization Committee of the German Industry*), February 1918, 39.

in the introduction after World War I of a normalized paper format (DIN) that, ignoring the unruliness of the English-speaking countries, is counted among the most successful worldwide standards.

Eyes That Don't See ... Communication

Concepts of movement and transformation such as industrialization, standardization, Taylorization, and rationalization played an all-important role in the architectural programs of the interwar period. Habitation broke away from its static foundations and shifted to a temporalized order, both physically and metaphysically—which could certainly not be described as (only) an effect of the architectural programs. Rather, these reacted conversely to this uprootedness that went along with the transformative processes within modern society and reformulated them into a program directed toward future developments.

If the classic house invented in the Neolithic age gave us the categories of "place, space and location," as German architectural and cultural theorist Hajo Eickhoff writes, in the modern city, these enclosing functions now pass to infrastructure.[16] Modern living, in which the house is reproduced serially in living units and detached from the floor, goes back, on the one hand, to the temporarily occupied cell.

However, this cell is more a part of infrastructures than of the house. Thus, on the other hand, the context of enclosure is dissolved into the network of infrastructures—mobile spatial cells (at least in fantasy), or living containers—on the circuits of the infrastructural networks. Eickhoff wrote, "If the original existence of humans was to live amid the absence of the house, today humans live in the cosmos of its emptiness, nestled in its technical universe.... The cosmic bonding of humankind between heaven and earth has given way to a dizzying technological dependency."[17]

Accordingly, Le Corbusier's *unité d'habitation* [unit of habitation] should be less a finished unit, a house, than an open aggregate, mechanically coupled and expandable, with the cell as a logistical box and a modular element for the production of space. The modern world and modern building should function according to the generalized principles of Ford's factory: serial production and optimized work flow.

This view was shared by Gropius, who, during the time that he was remaking the Weimar (and later Dessau) Bauhaus into a laboratory for researching new industrialized building techniques, was said to have stated on several occasions that he wanted to go down in history as the "Ford of housing." The restructuring of processes and architectures after the model of Ford's automobile production dominated the fantasies of planners from the most varied domains of society in the interwar years. The Ford factories in Detroit developed into a pilgrimage site of modernism in the 1920s. Into the 1950s, the specter of houses being produced and sold like automobiles haunted planning discourse. It was only then that the recognition began to spread that despite all the programmatic mobilizations of architecture, houses and automobiles were fundamentally different products.

In a series of programmatic articles in the journal *L'Esprit Nouveau* (published as a book entitled *Toward an Architecture* in 1923), Le Corbusier showed ocean liners, airplanes, and automobiles as models for a renewal of construction. In an infamous photomontage which could also be a caricature of the misanthropic megalomania of modernity, an ocean giant of the most current generation looms threateningly (to scale) over four gemsof Paris

"Eyes That Don't See ... Ocean Liners"—subheading from Le Corbusier's manifesto *Vers une architecture* (Paris: Éditions Vincent Fréal & Cie, 1966), 65. © F.L.C. / ADAGP, Paris / Artists Rights Society (ARS), New York 2014.

Disbelieving eyes: the futuristic appearance of the steamship in the city of yesterday. From Le Corbusier, *Vers une architecture*, 71. © F.L.C. / ADAGP, Paris / Artists Rights Society (ARS), New York 2014.

architecture. In this way, the *unité d'habitation* declared itself to be more a threat than a promise, but great honor was bestowed upon the steamship liner. It appeared as a new, moving monument of modernity against the immobile evidence of past building cultures. It became a spatial representation and a representative space of modern architecture.

One might also say that the imagescape, in which the modern programs for the industrial, modular reorganization of construction are reflected, is the mechanically mobile spatial unit. Machines and logistics were seen as a panacea, for which this period is also known (and decried) as the Machine Age.

Americanism: America versus "America!"

America, the technologically and economically dynamic young society in the New World, suffered from image neurosis and longed for the evolved culture and refined style of the Old World. Meanwhile, the melancholic, self-doubting Old World was afraid of losing its importance and of declining, and so it found salvation in the innovations of the New World.

This situation was characteristic of the relationship between the large continental European nations and the United States from the late nineteenth century well into the second half of the twentieth, and to some extent it remains to this day. It shapes the debate over new technologies, like the introduction of modern transport systems. And it has a particularly convoluted effect on the history of modern architecture, on the universalization of the international style of high-rise boxes of cells stacked on top of one another.

No names are so intensively bound up with the influence of America in 1910–1930 as those of ergonomist Frederick Winslow

Taylor and automotive entrepreneur Henry Ford. "Taylorism plus Fordism equals Americanism," reads the heading of American historian Thomas P. Hughes's remarks on the subject, in the formal style of Vladimir Lenin's famous formula "Socialism is electrification plus Soviet power."[18] The heading suggests how strong an influence the American rationalizers and system builders had, even on the development of the young Soviet Union and the doctrines of Leninism. Engineers of European and Soviet Russian society adopted from Taylor the belief, or were at least strengthened in it, that the principles of rational organization of work may be applied beneficially in all areas of society.

At the second International Congress of Modern Architecture, held in Frankfurt am Main in 1928 and attended by the entire European avant-garde of architecture and urban planning, a course was set that would be essential to the further development of modern architecture: the general recognition of the doctrine of the "apartment for minimal existence." Thus, Tayloristic principles were applied not only to building but also to the organization of living space itself—to life.

"While directly using Taylor's terms and methods," architectural historian Winfried Nerdinger states, "Gropius explained that one only needs a little space, provided that it is properly organized 'operationally.' In parallel to Tayloristic work, living should be dissected into its individual processes, organized into the most efficient sequence and spatially reduced."[19] It is the programmatic, constitutional generalization of the cell of minimal space as the basis of all building programs.

In the years that followed, there were a series of other such congresses. And it is an interesting coincidence—at least from a container-centered perspective—that the most important of these congresses, at which the Athens Charter was adopted, to which the grand city building programs throughout the world would refer in the following decades, took place in the same year (1933) that the foundation for international container transit was laid with the establishment of the Bureau International des Containers.

In 1932, American architectural historian Henry-Russell Hitchcock and architect Philip Johnson, a later student of Gropius, introduced the architecture of the European avant-garde to the United States through an exhibition at the Museum of Modern Art in New York. The show's title was "The International Style: Architecture Since 1922." What was perceived as the superiority of the United States in the 1920s—Taylorism, industrial mass production, and modularization—returned to the nation in the early 1930s as an aesthetic program. If European architectural modernity may be understood largely as a reaction to American technology, or its myth, the designation *international style* veiled this circumstance and, in this way, allowed it to become hegemonic.

When a large portion of the European architectural avant-garde and nearly all leading members of the Bauhaus emigrated to the United States after the Nazi seizure of power, the "white gods" arrived, as Tom Wolfe ironically described the scene.[20] (He plays on the fact that the box-shaped houses and spaces of classical modernity are preferably kept white.)

With their exhibition, Hitchcock and Johnson prepared the way. Soon thereafter, Gropius was appointed director of the architectural faculty at Harvard University, and Mies van der Rohe, the last director of the Bauhaus, became dean of the architectural faculty at the Armour Institute of Chicago (later renamed the Illinois Institute of Chicago). Within a few years, the American architectural scene, which had always tried to establish its own national style, changed completely. While the United States was on the verge of becoming the world's military and economic leader and far outpacing the Old World's cultural nations with its culture based on consumption and private property, in terms of both economic power and standard of living, architectural perspective was reformed according to antibourgeois, functionalist, and rationalist ideals that arose from the specific shortage situation in interwar and postwar Europe and from the image that Europe had created of America. Thus, the cell as the smallest unit of construction and the functionalist layout of gridded facades became architectural dogma in the United States.

In the meantime, however, a new era had dawned. Giant automobiles with shark fins, cartoon series with monsters and superheroes, oversized refrigerators, shopping centers as big as castles, sports stadiums for 100,000 people, and gigantic new suburbs characterized this new culture. Nevertheless, it was not the regionalism and eclecticism of the early high-rise years, nor was it the Prairie School or the Chicago School, that became the formative style of this American age. Rather, it was the international style by which "good" architecture was measured—according to Tom Wolfe's analysis, an aesthetics of "worker housing pitches up fifty stories high."[21] The international style formed the skylines of the American inner city and, since the United States was the leading nation of the Western world, the inner cities and new segments of the other Western capitals as well.

According to Wolfe, the primary symbol for this bizarre hegemony of the result of a misunderstanding is what became known as the Yale Box or the Mies Box, a smooth-walled, simple, box-shaped high-rise of glass, metal, and concrete, which would become the trademark of modern architecture for two decades:

At Yale ... [e]veryone designed the same ... box ... of glass and steel and concrete, with tiny beige bricks substituted occasionally. This became known as The Yale Box. Ironic drawings of The Yale Box began appearing on bulletin boards. "The Yale Box in the Mojave Desert"—and there would be a picture of The Yale Box out amid the sagebrush and the Joshua trees northeast of Palmdale, California. "The Yale Box Visits Winnie the Pooh"—and there would be a picture of the glass-and-steel cube up in a tree, the child's treehouse of the future. "The Yale Box Searches for Captain Nemo"—and there would be a picture of The Yale Box twenty thousand leagues under the sea with a periscope on top and a propeller in back. There was something gloriously nutty about this business of The Yale Box!—but nothing changed. Even in serious moments, nobody could design anything but Yale Boxes.[22]

French filmmaker and comedian Jacques Tati, who for most of his life was a Don Camillo–esque warrior against the hegemonic claims and demands of modernization, showed the Yale Box as a subtle decorative element in his film *Playtime*. The film, created between 1965 and 1967 with extremely expensive scenes built full size, can perhaps be seen as the ultimate film examination of life in serial box housing. It begins at that paradigmatic place—or, in keeping with French ethnologist Marc Augé, nonplace—of the uprooted stay in globalized interiors: at an airport. A long sequence of scenes follows in which Tati plays through possibilities of unease, slippage, nonarrival, and failure in the smooth box setting. Tati sets the resistance and humor of the anachronistic needs and body movements of his protagonist against a world of standardized veils for houses and people, of absolutely smooth materials and rituals without roots.

The Yale Box appears at a travel office or a kind of reception center, where a group of American tourists to Paris has just arrived. They don't see the actual Paris anywhere in the entire film. Modern Parisian life is acted out in an outer suburbia of supermodern buildings. On various posters hanging in front of the reception center's windows, the exact same skyscraper can be seen that is also in the background of the film as part of the urban scenery. The fact that the posters advertise different destinations—London, the United States, and Brazil—can be seen only by the changing name caption and by a few different symbolic local peculiarities marginalized by the broad high-rise, whether it be Big Ben and a double-decker bus for London or a Samba dancer and a palm tree for Brazil.

Little Boxes

In the United States, the local counterpart to the international high-rise is the serially produced single-family suburban home. It has also come under the scrutiny of urban sociological critique since the early 1960s. As the embodiment of standardization, social segregation, and dumb consumerism, it has become the symbol of what is bad in America for many, particularly for leftist critics. This criticism found expression in the 1962 folk song "Little Boxes," a hymn of the nascent American postwar protest culture. This song—written by Malvina Reynolds but later a hit recording by folksinger Pete Seeger—ironically takes as its theme the uniformity of the rows of houses and the conformity of those living in them and captures them in an image of hastily constructed, identical small boxes:

Little boxes on the hillside
Little boxes made of ticky tacky
Little boxes on the hillside
Little boxes all the same.

There's a green one and a pink one
And a blue one and a yellow one
And they're all made out of ticky tacky
And they all look just the same.

And the people in the houses
All go to the university
Where they were put in boxes
And they came out all the same.

And there's doctors and lawyers
And business executives
And they're all made out of ticky tacky
And they all look just the same.[23]

Box people in box houses. In their critics' eyes, suburban boxes and high-rise cells are similar—self-containers (Sloterdijk's term) for nuclear families and those living alone—in their function of surrounding the spatially and socially uprooted existence of modern subjects through architecture as the deployment of containers. From this perspective, the settlement of stacked containers, just like the trailer park of parked mobile living modules, seems to simply be a particularly radical realization of the program of mobilized living, a mirror for the truth of modern architecture.

The American suburbs of the 1950s first became the object of sociological investigation in William H. Whyte's *The Organization Man*, which studied a new type of midlevel employee, dressed in a uniform gray, all the men going to work on the same public transit system at the same time from identical suburbs. The proposals discussed in emerging urban and architectural criticism to combat this form of conformity were, and are to this day, the same as those levied by postmodern critiques of modern building—a proliferation and refinement of styles.

In contrast, American architect, designer, and architectural theorist Keller Easterling maintained in a series of writings from the late 1990s that in reality the dominant structure in the large suburban communities, the architectural principle, had almost

After the war, the workers of America preferred the little boxes to the big boxes. Aerial photograph of Levittown, Pennsylvania, one of the first suburbs of the new type for roughly 50,000 residents, the model for many similar settlements and a symbol of what its critics saw as the misguided city development policies of suburbia. Levittown was established in 1952, shortly after the completion of the first Levittown, in Long Island, New York, under the direction of William J. Levitt, who is frequently named as the founder of the modern American suburb. *Credit:* National Archives and Records Administration.

nothing to do with aesthetics. Rather, it was much more a matter of the *organization* of the building process, a logistics for the production of houses. In an essay entitled "Interchange and Container: The New Orgman," she wrote the following:

> The architecture of the mid-century suburb was organizational. It was not about the appearance of the individual house, but rather about the almost agricultural logistics for producing a series of identical building operations in succession, and it was prized primarily for the sheer numbers of houses it could produce. The organizational protocol was not merely that which facilitated architecture; it was the architecture.[24]

In the 1982 Czech children's cartoon film *The Mole in the City*, we see through the eyes of a little mole and his two friends, a rabbit and a hedgehog, as a forest becomes a modern planned city in a short period. Like a giant field, the surface is cleared, then leveled and arranged with houses and streets. It is a great mechanical movement that forms out of nature a second nature of concrete, from the forestry machines and earthmovers to the stacking of cells and the laying of street corridors. The urban traffic takes on the energy impulse of the cultivation and extends it into a kind of molecular particulate motion. Standardized mobile parts and standardized stationary parts together constitute the city system.

The depiction of the rise of a socialist planned city in the Czech cartoon and the analysis of the rise of an American suburban community by an American architectural theorist have in common the image of a quasi-agrarian, serial activity that revolves around not the aesthetics of the individual house but rather the rational production of the entire community according to certain defined procedures and forms.

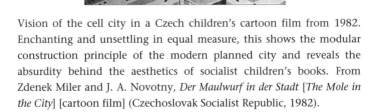

Vision of the cell city in a Czech children's cartoon film from 1982. Enchanting and unsettling in equal measure, this shows the modular construction principle of the modern planned city and reveals the absurdity behind the aesthetics of socialist children's books. From Zdenek Miler and J. A. Novotny, *Der Maulwurf in der Stadt* [*The Mole in the City*] [cartoon film] (Czechoslovak Socialist Republic, 1982).

A Dream Fulfilled: Living in Containers

Nowhere is the container's reputation worse than in architecture. After the failure of the high-flying, cell-based programs of modernity, the container must serve as a metaphor for everything bad, misanthropic, and technocratic. However, even after the (supposed) break with the dogmas of modernity, architecture continued to operate with cells. In a whole series of designs in the 1960s, architects attempted to make full use of the mobility of standardized room modules. These ranged from the unbuilt, trendy visions of the English group Archigram, which united the optimism and playfulness of an antiauthoritarian awakening with the technological utopia of the space age: multifunctional steel modules with rounded corners called *capsules* and *pods* by their designers that should be able to be inserted into or removed at will from an infrastructural network of steel tubes raised above the ground (the "plug-in" city) to Kisho Kurakawa's Nagakin Capsule Tower in Tokyo, presumably the most famous current monument to building with capsules. The idea behind this building was that in the future, when many such buildings stand in the center of the most important cities, the target group of overworked managers who work in a place only on a project-to-project basis can have their own capsules moved and inserted into another building.

With all the plans and visions that the famous representatives of the first generation of modern architecture developed for prefabricated modular building systems—whether Le Corbusier's Citrohan and Dom-ino systems, Walter Gropius and Konrad Wachsmann's packaged house or the experimental houses in the model Bauhaus settlement in Dessau-Törten, Buckminster Fuller's Dymaxion house and Wichita house, or Frank Lloyd Wright's

American system-built houses—this second wave of architectural designs shares in the fact that they are all realized in one or two prototypes, more often not even making it past the paper stage.

"The awful truth is that as industrial products, these houses were all either failures or non-starters," London architecture professor Colin Davies wrote in a 2005 book on the history of the prefabricated house.[25] According to Davies, the problem could lie in the fact that the father figures of modern architecture, with all their high-flying programs, did not build *houses* in which real people wanted to live and for which they were prepared to pay a price; rather, they primarily wanted to make *architecture*. In a pinch, a design would suffice as well. (This is a reassuring conclusion that the architects of the following modernist generations drew from the pretentious and unrealistic plans of their predecessors.)

In this context, it is not the case, as is claimed in many histories of architecture, that the idea of prefabricated, modular, and mobile living units had to fail because it came up against the fundamental, and ultimately irrevocable, static principles of the house. Since the beginning of the twentieth century, hundreds of thousands of industrially produced houses, mobile homes, and trailers have been sold in the United States alone. Firms like Perma-Bilt Homes, Home Building Corporation, National Homes Corporation, and Gunnison Housing Corporation usually offered a palette of models in more or less traditional styles and, as a rule, did not even have an architect in-house. Without bothering with a theory of the spatial cell or the proper human dimensions, these companies made their decisions simply according to the normative framework concerning aspects like street width, pallet dimensions, or material weight and in keeping with the demands of the future residents of their houses.

Page from a mail-order catalog of Sears, Roebuck and Company, 1910. The buyers can choose from different types of houses and pick from a model catalog of various doors, windows, fittings, and so on. © Sears, Roebuck and Company.

In the first half of the twentieth century, the retailer Sears, Roebuck and Company offered mail-order houses at a price of $650 to $2,200 that could be chosen and individually combined in a sample catalog. In the course of the roughly 40 years that the firm remained in this business, 450 different house models came to market, and more than 100,000 houses were sold. The bottom line is clear: the "normal" people for whom all the grand architectural programs were supposedly developed did not want architecture; they wanted a home. Whether that is a cell, a container, an elaborate system of wooden scaffolding, or a classical house of brick and mortar remained, at most, secondary. Consequently, they moved en masse into the new suburban settlements, while the ambitious social housing projects in the modern style were inhabited only by those who had no choice.

Industrial mass production and modern aesthetics generally coincide successfully, for residents, only when the extremely limited economic capacities of a pioneer or crisis situation restrict the facilities to a bare minimum. For instance, the first generations of the Israeli kibbutz structures in the 1940s and 1950s mostly consisted of nothing more than precast concrete shells. Or, to give a more recent example, the large housing construction program that postapartheid South Africa undertook for its largely impoverished populace also restricted itself to the rational building of the simplest box houses—at least by European standards. In 14 years, 2.7 million houses were built for more than 13 million people, according to the Ministry of Housing in early 2009. People were proud and happy to obtain these homes, despite the often great distances from their workplaces and despite the settlement structure of these townships, which are certainly reminiscent of American suburbs but which also continued the apartheid government's legacy of racial separation in living space.

Since the end of the 1990s, since standard shipping containers have been produced in such tremendous volumes and transported around the globe long enough that the first generation of them (after 10 to 20 years of service in transport) has been scrapped and they have been made available for other uses, a striking number of designs for container buildings has accumulated. Among the first to dedicate itself systematically to spatial constructions with containers was the New York–based office Lot-Ek, comprising Italian architects Ada Tolla and Giuseppe Lignano.

After having explored for some time the possibilities for reusing and misusing industrial and infrastructural components for architectural and artistic aims, in the mid-1990s they hit upon the container. The standard unit for their designs was the mobile dwelling unit (MDU), a standard shipping container fitted with all manner of drawerlike fixtures. American architect Aaron Betsky, curator of the 2008 Venice Architecture Biennale and specialist in temporary and conceptual architecture, said the following about the project:

Tolla and Lignano ... hit upon a building element that was in itself already part of the industrial world, and yet was meant to contain things. By importing it into the domestic sphere and cutting and pasting its various parts, they were able to turn it into a hybrid of architecture and technology. The container and the shipping vessel became the ultimate merger of everything architects were trying to achieve. It was a building block, an expression of systems, a moveable bit of a changing society, and something that could be found, rather than having to be constructed by using up resources.[26]

(a) In 2006, Zurich witnessed the new opening of the flagship store of the company Freitag, which makes bags from recycled truck tarpaulins. Erected next to railway facilities and an arterial road, the building consists of 17 scrapped shipping containers stacked on top of and beside one another. The company brochure says, "Thus arose a bonsai skyscraper in Zurich. Short enough to stay below the regulatory line for a high-rise, high enough to make even the most hard-boiled visitor dizzy." (b) © Alexander Klose.

Lot-Ek fulfilled all manner of smaller installation, addition, and renovation orders from private parties over the years with individual containers or container parts, and it brought about larger container installations in the context of art exhibits, but for the MDUs it developed visions of housing reminiscent of designs from the 1960s. On the margins of existing container system infrastructures, MDU ports were to be built, shelflike metaconstructions into which individual MDUs could be mounted and connected to the necessary utilities—a high-test hybrid of Le Corbusier's wine bottle idea from his *unité d'habitation* and Archigram's plug-in concept, only on the basis of fully developed transport infrastructure and its modified standard units.

In the architects' vision, the residential shelves would foster constantly changing neighborhoods as the living containers come and go. Whether such a living concept would generate interest remained to be seen. Nevertheless, it has been proven time and again that willingness to move is extremely low, even (and especially) for those facilities created for mobile and temporary living, like American trailer parks. However, the term *rack*, and the consistent organization not only of construction but also of living as a logistical operation, makes it clear that Lot-Ek is not afraid of the connotations of warehouse-style administration of living space. Rather, in light of this general finding about life in the twentieth and twenty-first centuries, it is rushing toward it headlong.

On March 19, 2002, a preliminary report appeared in the real estate section of the newspaper *Die Zeit*, beneath the somewhat sarcastic headline "A Dream Fulfilled: Living in Containers," about a planned container housing development in downtown London that was in the testing and advertisement phase. Since then, a whole array of such "normal" container housing projects

Design for a living container port from the architectural firm Lot-Ek. "Once the MDU reaches its destination, it is loaded into MDU Vertical Harbors located in all major metropolitan areas. The harbor is a multiple level steel rack, eight feet wide (the width of a container) and varying in length, depending on location.... A crane on rails travels the entire length, parallel to the structure. It grabs the incoming MDUs and loads them into spaces in the shelf." From Christopher Scoates, ed., *Lot-Ek: Mobile Dwelling Unit* (New York: Distributed Art Publishers, 2003), 58. © UC Regents, Art, Design & Architecture Museum (formerly University Art Museum), UC Santa Barbara, all rights reserved.

has been realized in various places around the world. The most prominent is probably Container City, a settlement of scrapped standard containers with studios and apartments, the first portions of which were built in the docklands of London in May 2001 and which were successively expanded. Since that time, they comprise more than 100 living or working units. The firm responsible, Urban Space Management, received orders from all parts of the world.

As early as 1998, Berlin architectural critic Dieter Hoffmann-Axthelm published a manifesto for the container as a unit of space for a new subsistence economy. This was supposed to be developed under the prevailing "large forms": the commercial regulations, building constraints, and restrictive lending practices that made the conditions for achieving independence much too high for a large portion of this nation's inhabitants, particularly immigrants. According to Hoffmann-Axthelm, the container was a "measure of the city in crisis" for economies "in the shadow of globalization." He continued:

The city in crisis is a city well-versed in small units and independent actors, the container the measure of the smallest concrete unit. What is exemplary is not the technical device, but rather the virtue of the demarcated area, the provisional settlement, the beginning, from which strong relationships can grow. Thus, the container can be rethought in any number of forms: a parcel of land, a house, a floor, a room—provided that it has to do with sizes that open up autonomy.[27]

The break with the ideals and paradigms of modern architecture and urban planning, or their revision, cannot be aesthetic. The cells are not the problem. The problem is the claim of rebuilding the world with them. On the other side is the pragmatism of the actual existing (former) transport containers. As a kind of nonarchitectural residence, to this day the container marks a zero point and consequently poses radical questions about architecture and the current state of the production of the built environment. What is a house? Where does architecture begin? What are the minimal and the most important requirements for a human residence? What is needed to transform a simple container into a home? And what becomes of the category "home" with such logistical container dwellings?

Darren Almond, *Mean Time*, 2000. An art container on a journey lets the system reveal itself. © Darren Almond, Matthew Marks Gallery.

8   Container World

People standing unsteady on rolling ground
Finally forced to see their existence with sober eyes.
—Bertolt Brecht

In the late summer of 2000, London artist Darren Almond embarked on a journey by container ship from the British Isles to New York to undertake his first solo exhibition. He was accompanying his most important exhibition piece, the work *Mean Time, 2000*: an orange 40-foot container into which an oversized digital clock with folding numbers had been built. The display stretched across the entire height and a good two-thirds of the width of one side of the container. The clock container sat atop one of the container stacks on deck, connected to the ship's own power supply. Since it was a radio-controlled clock that received signals at regular intervals from circulating communications satellites, throughout the entire journey it showed to the second the precise time for the zone from which the artwork was sent, Greenwich Mean Time, the prime meridian.

In a certain sense then, it was a smart box. One could see the whole thing as a parody of the vision (mentioned briefly in chapter 6) of making containers "intelligent" by means of RFID chips. In the exhibition at the Matthew Marks gallery in New York, entitled Transport Medium, the container claimed nearly the entire volume of the gallery's largest room. The atmosphere was filled by the constant clatter and hum of the mechanical display of the digits. The hard technical container artwork, with its aura of the industrial, its reference to the electronic timing of automated production processes, and its technoid sound, was contrasted in the exhibition by a series of five drawings, all entitled *Magnified System Diagram, 2000*, which Almond made during the five nights of his ship's passage.

These drawings were maps of the celestial constellations visible on each night. They stood in stark contrast to the image of the oversized clock, faithfully and accurately showing the time in the middle of the night, on an uninhabited island of industrial landscape at sea. On the one side, the clock showed the completeness and self-referentiality of a machine as part of a system in the second nature of global communications and transport networks. Their mechanics operated with perpetual precision. Nothing seemed to influence the mechanical course of things in this system, neither time of day nor year, neither storm nor calm.

Although everything is always in motion in the world of logistics, and the clock was relentlessly driving compliance with the timetable, it seemed that time stood still and history was at an end. On the other side stood a solitary man who gave temporal and spatial definition in the most archaic way, through the observation of the course of nature and the cosmos. In its extreme tension, Darren Almond's installation pointed to central questions of location, to the spatiotemporal constitution of the container system, its components and the things and beings that it included.

In addition to the reference to the standardized world time, the name *Mean Time* has another connotation: in the meantime.

The world in the container creates its own paradoxical meantime in the permanent stream. In his influential three-volume work *The Power of Identity: The Information Age*, Spanish sociologist Manuel Castells presents a comprehensive social-scientific examination of the past and present of our modern globalized world. With his juxtaposition of the "space of flows" of global information and transport technologies with the "space of places" of local markets and cultures, he provides a decision matrix to which many theories refer: "Space and time, the material foundations of human experience, have been transformed, as the space of flows dominates the space of places, and timeless time supersedes clock time of the industrial era."[1]

Perhaps the most radical change brought about by the container system is the broad dissolution of classical material storage and the mobilization of inventory in highly complex, precisely coordinated production processes, in which each part is delivered just in time. Silos, the historical antecedents to the modern warehouse, worked for thousands of years to establish places and cultures by being static containers of stockpiles that made the establishment of permanent human settlements possible. The more the stores are relocated onto the global conveyor belt, and the more the goods spend most of their time in the moving warehouse of the container, the more generalized is the new fathomless positioning in the space of flows, defined by the length of transport sequences.

Loss of Margins

"The world is always near us. No matter where" is the claim of the German container shipper Hapag-Lloyd. "All cargo, to all places, at all times" is the motto of the Zim Israel Navigation Company.

The list goes on. All globally operating transport companies make more or less identical promises. Underlying their offer of global availability at any time is the apparition of being able to eliminate spatial distances and locally distinct conditions completely in favor of a flat network of homogenously distributed locations that is subject only to the necessities of temporal organization.

In a 1949 lecture, the German philosopher Martin Heidegger questioned this vision of "planetary technology": "What is happening here when, as a result of the abolition of great distances, everything is equally far and equally near? What is this uniformity in which everything is neither far nor near—is, as it were, without distance?"[2] Heidegger feared that with the loss of distances and differences, there would be a loss of all possibility of intensive communality and local development—and with it, the loss of what had presumably been the motor for individual and cultural evolution from the beginning.

However justified such concerns might be, one must also guard against confusing the company slogans with reality and see them as what they are: marketing gimmicks of an ideology of technical progress and omnipresence in the (supposedly) post-ideological sphere of free competition in the capitalist world system. Transportation's economic ideal of around-the-clock availability is a fantasy. It evens out complexities and ignores deviations from the ideal model of distribution, whether it be disturbances inherent to the system, such as traffic jams caused by the congestion of transport infrastructures; shortages (of fuel, containers, or berths in the harbors that are constantly at the upper limit of their capacity because of continuous expansion); local resistance, like pirates or labor organizations; adverse environmental conditions; or the fact that a large part of the world,

Advertisement for a container repair company, 2006. The slogans of international companies in the logistics industry spark a belief in complete worldwide globalization while also feeding off it.

especially in the Southern Hemisphere, is only very sparsely covered by the container system.

What is specific to the global transport logistics system and significant about a theory of the status quo of the globalized world is that it unites the time- and space-leveling logic of the space of flows with the local spatialities and temporalities of its material elements. Containerization is a latecomer in the industrialization process. The container transport system consists of enormous, heavy machines. It moves hundreds of thousands of tons of goods and dead weight with precise and complex synchronization of multiple processes.

In all of its clocked consistency, external circumstances like distances, day and night, or streams and winds certainly play a role—how could it be otherwise? And it is not just in the banal factor of fuel costs that this is expressed. There is also the influence of dockworkers, labor provisions, and local regulations; national and international agreements on tariffs, subsidies and taxes; the treatment of refugees and terrorists; raw material capacities and prices; and the local state of development of the respective technologies. If one reduces these influences to the status of mere disruptions in a system that is actually raised above all inner logic of local conditions and material configurations, then one fails to recognize that the system is and must continually be assembled from all these components.

Thus, for instance, the reason the grand history of the port of London—the world's largest throughout the nineteenth century and into the twentieth—ends abruptly with containerization does not lie simply in the insufficient depth of the Thames riverbed and the lack of space for containers in the London metropolitan area. Rather, the Port of London Authority's plan to establish the port of Tilbury at the mouth of the Thames as a new, upstream site of the port of London for container shipping failed because the dockworkers' union, still powerful at that time, paralyzed the port through a strike in the decisive phase of 1968–1970.

Thus, as the course of transatlantic traffic was set, the first container ships sidestepped to the previously unimportant port of Felixstowe, which quickly became Britain's most important container port and which retains that position to this day. In the complexities that constitute the reality of the logistics system, and with it globalization, the logics of the container play their specific role—the logics of temporary, sequential, and serial inclusion and exclusion and the supply of standardized room units.

This chapter is concerned with this space-time regimen of the container, with the relationship between the container principle and its character as a commodity, with containers for people, and with what happens with the things, the people, and the organic substance in containers. In the process, a few things that have already been discussed or at least touched upon in previous chapters reappear in a series of scenes from which one may perhaps piece together what constitutes the container world and what life in it is like.

Naked in the Container

As part of the Vienna Festival in June 2000, the director and performance artist Christoph Schlingensief (who died an untimely death from lung cancer in 2010) erected a deportation camp of containers on the Opernplatz. It remained there for a week. There were 12 refugees living in it who had applied for asylum in Austria. Every day two of them were chosen from the camp;

promptly at 8 p.m. came the "deportation." One could follow the lives of the refugees in the container on the Internet through six webcams and also take part in the deportation vote.

The whole matter was a reaction to the increasingly restrictive refugee policy of the European Union (EU) states and to the fact that since the beginning of the year, a right-wing political party was cogoverning with the Freedom Party of Austria under Jörg Haider that waged its election campaign with xenophobic, populist slogans. *Ausländer raus* ("Foreigners out") was emblazoned on a large panel on the container colony's roof. Excerpts from Haider's political diatribes ran from a tape. Schlingensief chanted the slogan "Foreigners out!" over a megaphone to the mass of people on the Opernplatz in the style of an agent provocateur. More than a few of the spectators agreed.

Schlingensief's container was also a bitter parody of the inhumane mode of operation of reality TV shows like *Big Brother*. The reality TV format, first introduced in the Netherlands in the 1990s and then in Germany, was as controversial as it was successful, and it had imitators the world over. In this kind of program, the viewers take part in the private lives of more or less normal, everyday people who find themselves under tightened experimental conditions, and the viewers can crown one of these people as the chosen one by gradually voting others off the show.

This format is widely known in Germany as a "container show," since the original broadcast of *Big Brother*, screened from a remote studio of the station RTL in a suburb of Cologne, was produced in living containers. The show's great popularity and its high quotient of scandal led to regular reporting on the program in the news media. The word *container* has expanded its scope of meaning and no longer stands merely for reality shows

A  B

C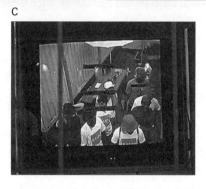

Schlingensief's controversial asylum container show on Vienna's Opernplatz in summer 2001. It was a beacon against the rightward shift in the Austrian government under Haider and against restrictive foreign policy of "Fortress Europe." Every day, thousands came to the container camp, discussing and shouting. The project website, through which one could select and "deport" the temporary residents of the container installation, had more than a million visitors during the one-week action. Stills from Paul Poet's documentary film *Ausländer Raus! Schlingensief's Container* [Foreigners out! Schlingensief's container] (Vienna: Bonusfilm, 2005). © Bonusfilm.

but also refers, rather ironically or critically, to the whole system of the media machinery, based on spectacle and the exploitation of the private.

Phrases like "a continent in a container" and "multicultural container" were tossed about in the July 2003 newspapers *Süddeutsche Zeitung* and *Tagesspiegel* in reference to the African edition of *Big Brother* airing on South African television. However, another typical formulation came in a February 2005 article in *Süddeutsche Zeitung* about former ski jumper and superstar Sven Hannawald, who ended his career after a mental collapse because he could apparently no longer handle the pressure of publicity produced by the media machine. "Eventually, he had to break free from the container," the article said.

The container event, which counted among Schlingensief's best and most impressive works, caused a huge uproar and tremendously emotional reactions. In 2005, the very worthwhile documentary film *Foreigners Out! Schlingensief's Container* by Paul Poet appeared, in which all of these voices were captured and the course of the entire event was brought into sharp focus.

The things that characterize the modern, globalized world, particularly after the end of the Cold War power blocs, are together in the film. On the one hand, there are open borders, global communication networks and free trade, and the trafficking of goods, information, and people to an extent never before seen. On the other hand, there is a fortresslike closing off of the rich portions of the world to the (unregulated) influx of people from the poor and war-torn regions; there is also a more than unjust distribution of access to resources, from basic foodstuffs to education and information to work and prosperity.

A general mobilization and dissolution of regional living conditions that had already begun with industrialization in the nineteenth century expanded its sphere of influence with renewed vigor to the farthest reaches of the world in the late twentieth century, and a large-scale regulation and slowing of these movements was triggered by uprooting.

This is the tiered "container order" in the affluent societies of Europe. The little people seek their fifteen minutes of fame in the TV "container" or on TV shows that work according to principles similar to *American Idol*. In exchange, they put their dignity and their right to privacy at stake. Expensive inner-city plots are being filled with cheap temporary housing quarters of stacked containers. Construction workers live in them, many of whom have no work permit and are poorly paid. Even so, this work offers them the chance for a livelihood in the poorer countries from which they come. However, the most underprivileged of the global distribution system—the refugees without money or papers and without the prospect of returning to their countries of origin—are forced into container camps or similar internment and regulatory architectures.

These refugee and deportation camps, mostly situated in close proximity to borders, maintain a permanent state of exception. Confined there are people whose only offense is trying to gain entry to the rich part of the world without the proper permission, or who have no papers at all that designate them as a citizen of any country. They are confined often for up to 12 months or even longer, until their asylum application has run through the bureaucratic entities—in most cases with negative results.

Italian philosopher Giorgio Agamben writes about the legal reality of refugee camps, in which the inmates are reduced to "bare life," whether in totalitarian or democratic states: "In all these cases, an apparently innocuous space ... actually delimits a space in which the normal order is de facto suspended and whether or not atrocities are committed depends not on law but on the civility and the ethical sense of the police who temporarily act as sovereign."[3]

It seems logical to compare the state of exception of naked people' in containers—stowaways, interned refugees—with that of goods. Stripped of their civil rights (or never endowed with them) and reduced to the status of bare life, the people in the containers reach a level of simple things (like house pets or slaves).

"You are now the property of the United States," Jamal al-Harith, a British citizen and a former inmate of Guantanamo, was told upon his arrival at the U.S. prison camp in Cuba, as he told the audience at a London conference in November 2005. The infamous camp was established by the George W. Bush administration after the attack of September 11, 2001, and his successor, President Barack Obama, hasn't been able to get the cooperation of Congress to shut it down, despite his campaign promise to do so. Guantanamo has become a "political myth of the early [twenty-first] century," journalist Roger Willemsen wrote in the preface to his series of interviews conducted with former camp detainees.[4]

The fact that containers would come into use in this space, conceived from the beginning as a legal vacuum, is no wonder—they belong (along with barbed-wire fences, barricades, barracks and prefabricated concrete cells) to the standard repertoire of modern police-state and biopolitical governance. (In the case of Guantanamo, one would have to add the metal cages where the detainees of Camp X-Ray were kept outdoors until it was closed in May 2002 as a result of international outrage.)

These measures for organizing space justified by a crisis or a declared state of exception should be able to be disassembled as

Container World

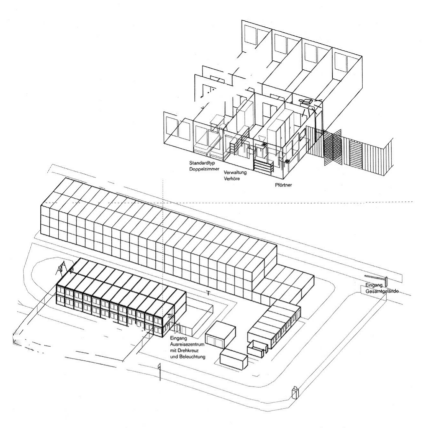

Spatial representation of the repatriation center and asylum seekers' home in Fürth, a container camp for refugees who cannot be deported because of a lack of papers. "Here, asylum seekers are accommodated who are accused by the authorities of concealing their heritage and not cooperating in obtaining a passport.... In the camps, the migrants are to be worked on until they leave 'voluntarily,' go underground as illegals or, after clarification of their citizenship, can be deported regularly. However, in contrast to a deportation hold, placement in the repatriation centers is not time-limited." From "Geography of the Fürth Repatriation Center," *An Architektur*, May 2004. © An Architektur. [Top: Standard double room—Administration, Hearings—Gatekeeper; bottom: Entrance to the compound.]

quickly as they can be put up. German cultural scientists, artists, and art historians Axel Doßmann, Jan Wenzel, and Kai Wenzel conducted a comprehensive study of the relation between statehood and temporary architecture and summarized it with the maxim "be faster": "faster than other armies, other states, other companies."[5]

In the same volume, media and cultural theorists Tom Holert and Mark Terkessidis explain why containers lend themselves to this kind of simultaneously accelerating and decelerating spatial regime:

Where a container sits, or several are stacked on top of or lined up beside one another, specific circumstances immediately dominate, in which the lasting and the fleeting enter into an indeterminate relationship. Containers create temporary places, because they already embody a convention of the temporary. In this respect, one could say that containers make it easier to recognize the temporary and use it.[6]

Once again it is a matter of specific forms of politics with images. The container settlement for foreign construction workers or asylum seekers says to the neighborhoods of citizens, who find themselves mostly unwillingly confronted with this situation, "I'll be gone soon! Don't worry, this is just a temporary installation."

Also, brutish, highly technological border fortifications erected at a tremendous expense serve to present an image of security. However, closer inspection often reveals that the technological and financial investment bears no relationship to the actual flow of refugees—and that the flow of illegal labor establishes itself beside or underneath the border regime, because branches of the local economy depend on it.

A prime European example of such an economy would be the cheap vegetables from the Canary Islands and southern Spain, which people in central Europe have benefited from for several years as an alternative to Dutch greenhouse products. Entire regions that used to be structurally weak and poor have converted to tomato cultivation and become rich areas. Without the inexpensive labor of illegal migrants (and without the infinitesimally low transport costs in the container system), there would be neither this new prosperity nor cheap tomatoes. A prime U.S. example is the important economic role of illegal immigrants from Mexico and Central America coming over the U.S.-Mexican border, as fortresslike as it may be.

Amid all this—and one should spell this out so as not to make a construction form responsible for circumstances that result from social organization—containers are the more dignified variety of repository. A large portion of the millions of people housed in refugee camps are forced to live in tents or structures cobbled together from cardboard, corrugated iron, or other waste. Even in the United States—the land of mobile homes and trailer parks—many people have lost their homes as a result of the financial crisis of 2008 and have been forced to live in tent cities on the margins of metropolitan regions.

Stowaways and Other Undesirable Cargo

As seen through the container, the drama of global economic inequality has two continually recurrent scenes. The first scene is of container ships sailing on the "ocean highways," along the coasts of the poor nations and continents, but not docking in port. The two most important shipping channels, the Suez and Panama Canals, help the majority of ships to avoid having to pass through the Southern Hemisphere at all.

In this respect, then, the return of pirates noted since the 1990s and their alarming increase in activity might also be seen as an attempt to redirect some of the world's mobilized wealth into its poorest parts.

The second scene is of containers that are opened in the ports or other transfer stations in rich areas of the world, when stowaways appear—dead or alive. Stowaways are people who are determined or desperate enough to attempt the dangerous journey packed into a steel box, and they often don't even know where the ship is going or how long it will be under way.

In a German Television Report from 2004 entitled *Tapping in the Container*, one can see how such people are processed in the port of Hamburg. While they are still on board the ship, a decree of rejection is issued and pronounced, a legal trick that ensures that the refugees go on land but legally never touch German soil. For identification, examination of their status, and administration of a standardized interview officially entitled "Questioning of Stowaways," they are brought into Department 2 of the Hamburg harbor police station, which is, incidentally, an extremely prosaic building of prefabricated cells, as if to underscore the fact that a refugee is entering a continuum of container spaces. From here the refugees are immediately rejected in most cases and sent back to the nations from which they departed.

The shipping company whose ship transported the refugees without the proper papers is responsible for the process and expenses of the return transport. This rule is valid for three years from the date of entry—in other words, even if an asylum application is filed and rejected after completion of the usual process. And it is in force in all EU member states that are parties to the Schengen Agreement as well as in the United States, Canada, and Australia. This generates a cost pressure on the part of the

Container World

Globalisierung in voller Fahrt — SZ-Zeichnung: Murschetz

"Globalization at full speed," reads the caption of this cartoon by Luis Murschetz. Floating container fortresses circle the globe with luxury goods. They brought nothing for the inhabitants of the poor regions. © Murschetz, *Süddeutsche Zeitung*, Jan. 27–28, 2007.

shipping companies and ensures that captains and crew act in the isolationist interests of the rich nations.

Reimer Dohrn, of the working group Stowaways in Hamburg Harbor, explained, "The shipping companies face a problem. They are obligated to organize the return travel. They delegate this problem to ship insurers. And, in cooperation with the harbor police, they handle the stowaway problem in the fastest and cheapest way."[7]

Two stowaways who were sent back on the same ship on which they arrived after a brief stay on land of a few hours were able to be interviewed by the director of the ARD-TV report, Manfred Studer, in their place of departure in Lagos, Nigeria.

"We knew it would be very hard and dangerous," one of them said. "But we were so desperate. We wanted to get away, out of Nigeria and into Europe."[8] And they left no doubt that they would try it again, even if it could cost them their lives.

English director Michael Winterbottom shows how that can happen from an internal perspective in a documentary fiction film entitled *In This World*. The film, awarded the Golden Bear at the 2003 Berlin International Film Festival, portrays the fate of two refugees who set out from Afghanistan for England. (Apart from Ireland, Britain was at that time the only EU nation without immigration restrictions.) Locked with an Iranian family in a small steel box placed inside a truck's trailer, all the refugee container's occupants suffocated on the way from Istanbul to Trieste, with the exception of a baby and the younger of the two refugees, who ultimately made it to London.

In order to help avoid such terrible fates, which happen again and again, and to raise awareness of the situation of eastern European refugees, Romanian artist and curator Matei Bejenaru made a travel guide in early 2005 with a bit of sarcasm. It shows Romanian emigrants who have no prospect of getting the proper visa for a legal entry the illegal paths to England and Ireland, gives tips on where they can find like-minded people in the French and Belgian port cities, and advises how they can counteract the dangers of hiding in a truck or container. The problem of legal entry was solved for Romanians with their nation's entry to the European Union on January 1, 2007. However, for refugees from other nations, the tips that Bejenaru gives could still be of use:

> The containers can be easily unlocked with a metal lever. If you are handy, you can unlock it without breaking the seal; thus the breaking in could be observed only at a close look, which doesn't happen when the container is loaded by the crane.

Container World

> There is a panel on each container stored at the harbour, with all the information about the destination, departure and the content. Along with the information from the harbour newspaper, you have all the necessary data not to miss your destination. Once in the container with the desired destination, check the ratio of the amount of load to the remaining space. The more space you have in the container[,] the easier you can breathe. Each container has two little air holes on the upper part of the door side.... It is advisable that no more than three persons should be in the container.⁹

Refugees are only one element in the new repertoire of undesirable cargo, however. With the changing view of the world after the attack of 9/11, the view of containers changed as well. As icons of globalization, they no longer stand only for the omnipresence of the capitalist system but now also stand for the dangers and fears that the worldwide consolidation of commercial and cultural spaces conjures up. Containers are x-rayed and unsealed or broken open, and all of globalization's evils come to light.

Roberto Saviano's 2006 report *Gomorrah* also begins with a container scene. The southern Italian author opens his account of the machinations of the Neapolitan Mafia, the success of which has forced him to live under constant death threats, with a memorable image that speaks to the frightening side of the worldwide distribution of production sites and labor. When a container being loaded in the port of Naples bursts open, the frozen bodies of Chinese migrant workers fall out:

> The container swayed as the crane hoisted it onto the ship. The spreader, which hooks the container to the crane, was unable to control its movement, so it seemed to float in the air. The hatches, which had been improperly closed, suddenly sprang open, and dozens of bodies started raining down. They looked like mannequins. But when they hit the ground, their heads split open, as if their skulls were real. And they were.

Men, women, even a few children, came tumbling out of the container. All dead. Frozen, stacked one on top of another, packed like sardines. These were the Chinese who never die. The eternal ones, who trade identity papers among themselves. So this is where they'd ended up.[10]

The passage goes on to say that there were the "wildest fantasies" circulating about the destination of these bodies in the perpetual stream of Chinese workers, that they would be cooked in restaurants or thrown into the crater of Vesuvius. Instead, they had had money withheld from their salaries to be sent back in the event of their deaths, for "a space in a container and a hole in some strip of Chinese soil."[11]

In this story, the container becomes a somewhat unearthly place of disappearance into a superhuman context. Is there a logistics in the service of souls? What happens to the immortal components of a person whose body is deep-frozen and stacked in a container with a mass of other bodies like a hunk of meat and transported to its final resting place? With the project Mission Eternity, running since 2006, the members of the artists' group etoy, network art pioneers who sought out the container in the late 1990s as a logical extension of their artistic access to a world of global networking, explore the idea of an immaterial, immortal remnant in a world of nearly unlimited data production and circulation. They outfitted a container as a multimedia memorial space, traveling around the world to festivals, biennials, and other artistic or cultural events.

In this multimedia piece, *Sarcophagus Tank,* pictures, texts, and sounds with remembrances of the departed can be downloaded and reproduced. The records are called Arcanum Capsules. They are based on XML-files (this is the format on which relational databases are built as well) that replicate with the help of software called Angel Application, according to the principle of ocean storage. These data packets are everywhere and nowhere, permanently circulating around the globe from memory to memory in a network of database servers and mobile communications devices.

However, by entrusting the question of immortality to temporal logistics, the etoy members place themselves firmly on the operational side. The fact that eternal rest becomes permanent commotion and the final journey becomes permanent circulation—if one wanted to be mischievous, one could say something about electronic soul processing—does not seem to cause them any problems. In this way, they skip past the scandalous metaphysical core of the container logistics system.

Peter Sloterdijk points out that since the beginning of the modern era, since the lessening of belief in a divinely inhabited heaven into which Earth is caringly admitted, people know "that they are somehow contained or lost—which means nearly the same thing now—in the boundlessness."[12]

Even and especially if global, satellite-assisted logistics promises a new preservation in seamless localization, the scandal of modern humanity's mere containment (in contrast to being fully present and secure), the complementary counterpart to which is emptying, is expressed nowhere more clearly than in the example of the container, culminating in the definition of being "both container and emptier," as philosopher Hannes Böhringer writes. "It actually contains nothing. It is into this nothing that the cargo is thrown."[13]

Pandora's Steel Box

The principle of container transport and the reason for its sweeping success lie in conceptualizing the container as a metacontainer whose cargo is *not* of concern, as long as you do not find yourself at the beginning or the end of the transport chain. The container functions as a kind of black box of transport. As explained in chapter 6, the black box was a model developed within cybernetics to deal with machines whose functionality was not clear.

Sociologist Niklas Luhmann found this principle replicated in dealing with people as well. Communication and mutual understanding could be possible only on the basis of a black-box process, in which each person would have to accept that he or she was understood by the other without obtaining any certainty about the validity (or falsity) of this presupposition. Luhman extended this finding to all social contexts, or "social systems," in his terminology. These were also formed through the interaction of "highly complex meaning-using systems that are opaque and incalculable to one another."[14]

In such a world, in which communication and community building present themselves as reciprocal projections of mutually inaccessible and closed systems as an effect of the interaction of black boxes, the container—opaque, sealed, and system building—becomes an allegory of the social. However, what happens when the formerly neutral containers are filled once again with meaning—from the outside?

"There will always be more things in a closed, than in an open, box," French phenomenologist and philosopher of science Gaston Bachelard noted as a kind of poetic container rule in a painstaking study of domestic spatial forms first published in English in 1964. "To verify images kills them, and it is always more enriching to *imagine* than to *experience*," he wrote.[15]

As the null point of meaning, the universal container represents an ideal medium for speculation and presumption. For a long time, the recurrent motif of the container in the systemic network was prevalent in the container stories—a kind of flow chart of transport with the container as an image of itself and simultaneously as a symbol of the overarching structure carried by it—an icon of logistics.

Recently, however, the container has increasingly become an object on which to project other meanings. In a modern retelling of the myth of Pandora's box, the shipping container is broadly subordinated to containing evil things. The containers have been transformed from black boxes of globalization into Pandora's steel boxes. Under the spell of this the most famous of all container myths, does even the constitutively emptied transport container refill itself with metaphysical contents?

According to the ancient story—in the form in which it has entered modern bourgeois culture since Erasmus of Rotterdam—the gods took vengeance for the sacrilege that Prometheus committed by revealing to humans the secret of fire: the energy source and thus the prerequisite for almost all technology. They sent an irresistibly beautiful woman named Pandora, who was supposed to bring Prometheus a container with all the world's evil inside it. (Incidentally, in the original Greek texts, the object is a *pithos*, a large storage jar; only in the modern tradition did this become the proverbial *pyxis*, or "box.") It was not Prometheus, however, but his gullible brother Epimetheus, who received the gods' gift—against the advice of his brother. At the first available opportunity, Pandora, who became Epimetheus's new wife in the meantime, opened the vengeance present, and humanity was afflicted with all evils and vices: mortality, jealousy, resentment, and blood lust, which persist to this day.

In the summer of 2007, American television aired an ad campaign that dramatically staged the simultaneous simplicity and enormity of the Pandora scenario, which spread throughout the discourse on national security within the rich industrialized nations after the attacks of September 11. Supported by threatening music, it opened with a line, spoken as well as written on the otherwise black screen: "Since 9/11 it is one of the greatest threats we face ... " Subsequently, a triptych of terror assembled on the screen. In the middle of the divided screen was the image of a mushroom cloud; on the left side, a still photo of Osama bin Laden; and on the right side, a photo of a container ship in port. The voice-over continued, " ... a nuclear weapon, in the hands of Osama bin Laden, shipped through an American port."

This TV spot was part of an ad campaign organized and financed by Wake-Up Wal-Mart, a network of American political activists. Standing behind it was the American trade union United Food and Commercial Workers, which struggles particularly against the employee policies of Wal-Mart, the nation's largest employer. Supported by more pictures of a container port, the inside of a container loaded with material labeled as radioactive, and terrorists marching in the desert with gas masks and machine guns, the voice-over continued, "But even though a Wal-Mart container arrives in the U.S. every 45 seconds—containers that could carry the weapons used in the next terrorist attack—Wal-Mart opposes scanning 100 percent of port containers, leaving America vulnerable to protect their profits."

It is not only the exaggerated ambition of a few marketing strategists trying to cook up a drastic campaign against a more powerful opponent that produces and exploits such an ultimate catastrophic fantasy. The strangest coalitions come about under the spell of threat myths. In defending its package of measures for homeland security, the U.S. government itself promoted the nightmare image of a bomb in a container. In 2002, under the name Container Security Initiative (CSI), it started a global campaign for the control of container traffic. A primary result was the legislative package Public Law 110-53, adopted in August 2007, to increase the security of international container ship transport and to protect the homeland according to the recommendations of the 9/11 Commission Act.

The "specter of a nuclear bomb, hidden in a cargo container, detonating in an American port" led politicians to vote for the law, as noted in an Associated Press report of July 22, 2007. It represented a drastic interference in the well-established processes of global container transport. One of its central clauses stated that by 2012, all containers destined for import to the United States must be inspected through nonintrusive technologies (i.e., gamma or X-ray image and radiation detection) before being loaded in foreign ports. At this time, this affected up to 700 ports with connections to U.S. ports that would have to be outfitted with the new technology in order to examine the 11 million containers that are currently transported into the United States annually (a goal that still has not been fully attained by 2014).

The CSI program strives for an expansion of American homeland security measures to hubs throughout the entire global network. Each port is supposed to become an exclave of the secured American homeland, including American CSI inspectors; the international shipping routes are to become homeland corridors. The 24-hour advance manifest rule, which stipulates that the manifest of cargo loads must be submitted electronically 24 hours before the arrival of a ship into an American harbor, is in effect today.

As is common in the international passenger control system, the basic attitude toward legal access has reversed in dealing with transport boxes. Every container is suspicious, presumed guilty rather than innocent. The data are subjected to a review called a screening. If a conspicuous profile is found because a container comes from a country classified as problematic, is packed by a company not viewed as trustworthy, or contains goods regarded as suspicious according to the manifest, then the container is examined more closely in port. The examination has two steps: the primary inspection, through X-ray or gamma ray screening of the container to determine its contents; and the secondary inspection, which is carried out by a team of inspectors who open and search the container if their suspicions that dangerous or forbidden materials are on board are corroborated by the scan.

The aspiration of subjecting each of the millions of containers that are imported into the United States to a thorough investigative process seems unrealistic. Even with the primary inspection, no more than 20 containers on average can be processed each hour by each scanning system. Indeed, at the height of the current state of technology, scanning a container takes no longer than 30 seconds, but the evaluation of the images by the people behind the screens can take up to 15 minutes.

The length of a manual search of a container in the secondary inspection can range from four hours for 15 to 20 inspectors to three days for 5 inspectors, as shown in a 2006 study by the Rand Corporation (itself above any suspicion of having a lax or even "subversive" posture toward U.S. security interests). As the law threatened to massively slow the process of global goods transport and intervene without any consultation in the sovereign powers of other nations, the resistance from both government and industry has been so great, internationally as well as within the United States, that we can hardly assume the law will be realized in this form.

Indeed, its implementation was delayed for two years by Secretary of Homeland Security Janet Napolitano in summer 2012. She called it impracticable at present and said that it would have a negative effect on the global flow of goods, and she even tried to convince Congress to eliminate the mandate.

We may also speculate that the politics of power and economics were behind this super signifier of the threatened homeland, since the United States has lost its leading position in international shipping. Thus, the CSI could also be interpreted as an attempt to get a foot back in the door of global cargo transport. There is also the obvious suspicion that the initiative was not least a symbolic measure intended to demonstrate strength and decisiveness to the public. After the 9/11 attack, a flood of public relations actions and cultural productions of all kinds invoked the myth of the birth of a new America from the rubble of the World Trade Center simply through the constant repetition of the motif of the collapsed towers. Images from politics, advertisements, and entertainment reacted with one another, emphasizing and strengthening one another.

During the Cold War, it was the total-catastrophe scenario of atomic bombs exploding simultaneously over cities that constructed the reality of the atomic threat as imagination, as Jacques Derrida worked out in "No Apocalypse, Not Now," his 1984 analysis of the reality of the nuclear conflict. Nowhere is the bombing "theater," the image of a bomb that destroys the whole Earth, staged more convincingly than in Stanley Kubrick's 1964 film *Dr. Strangelove or: How I Learned to Stop Worrying and Love the Bomb.*

When the bomb was lost, or at least suffered a massive reduction in importance as a guiding and power-stabilizing factor, with the fall of the Iron Curtain, a new motivator fed by fantasies and fears had to be found. At the beginning of the twenty-first century, a decade after the dissolution of the Soviet Union and with the 9/11 attack, this proved to be the "war on terror," no more winnable than an atomic war. The acclimatization engendered during the long years of the nuclear threat to the negative reality of an event produced solely through a deterrent image found its continuation in the invocation of the scenario of the terrorist "WMD" (weapons of mass destruction) in the container. A series of theatrical and television films have already played this out. And we have learned to love the bombs—the diffuse, ever-present threat through the images of terrorism, among the most powerful of which is the notion of a container filled with a nuclear bomb.

Container Revolution

Peter Sloterdijk wrote, about the modern subject, that "he who goes into the world like a bullet into battle," who concentrates on the future as project, "needs a gun from which to be fired." According to Heinrich Mann, Sloterdijk continues, for Napoleon, whose conquests mark the beginning of modernity in many places, this "artillery" was "the revolution," "by which the epitome of offensive missions is designated, through which the messianic radicals since 1789 see themselves positioned in its categorical 'onwards.'"[16] What artillery is responsible for the deployment of standardized containers on their offensive missions, which had and have a part in fundamental life changes equal to political revolutions?

Today, "messianic radicals" can be found primarily among technicians and mostly in the service of large companies. Regardless of how superficial the permanent technological changes are and how transparent the marketing behind the proclaimed "revolutions" are, this is the mode into which we have moved and will presumably continue to move, relentlessly, for the foreseeable future. Even, and particularly, if the technologies seem unspectacular and not worthy of further discussion, a second or third look at their development and their significance can often be very revealing.

Thus, the English soldier T. E. Lawrence ("Lawrence of Arabia") was said to have remarked that the introduction of the tin can in the first half of the nineteenth century "had modified land war more profoundly than the invention of gunpowder."[17]

We naturalize the inconspicuous technologies most indiscriminately and passively. For example, we have long since grown accustomed to finding nearly all objects of daily use in standardized packaging. The days are gone when local "ma and pa" stores offer a small range of goods and unpackaged products that are weighed out according to the needs of the customer. However, the broad implementation of industrial packaging in the industrial nations of Europe began only in the 1950s.

Until the middle of the twentieth century, a large portion of goods—like flour, rice, sugar, meal, or legumes but also cookies, bonbons, and even cigarettes—were sold loosely. It was only the introduction of the principle of self-service and of the supermarket that allowed packaging to become predominant for goods. Whereas the first supermarkets opened in New York in the 1930s, Germany had none until after World War II.

Historiography has to thank the architectural theorist Siegfried Giedion and his 1948 book *Mechanization Takes Command* for his attention to such details and their link to larger and more fundamental social contexts. His programmatic project of an "anonymous history" of the "outwardly modest things" aimed to work out the "fundamental attitudes to the world," as materialized in certain objects and artifacts, and "these attitudes set the course followed by thought and action."[18]

In so doing, he examined and linked developments in technology and everyday life that had not yet been brought together. For instance, he was able to show that conveyor belts were not first used in the industrial montage of dead matter in the Ford automotive factories in Detroit but had been employed a good 40 years earlier in the serial transformation of living material into standardized food in the early industrial slaughterhouses of Cincinnati. The principle of the modern conveyor belt is brought to bear in a particular phase of the slaughtering process. After being captured, killed, blanched, and depilated, the carcasses are hung on a suspended railway, two feet apart, and moved along before a series of workers. An article in *Harper's Weekly* on September 6, 1873, described the process: "one man splits the animal, the next takes out the entrails, the third removes heart, liver etc. and the carcass is washed by the hose-man."[19]

More pointedly, the serial animation of things in the industrial montage is historically and systematically preceded by the serial killing of animate beings in the industrial processing of foodstuffs. From this perspective, containerization is the logical continuation of a key strand of mechanization. In order to be able to process material efficiently, standardized sizes and forms are necessary. Where this is not the case from the beginning, as with industrial montage, which can customize its machines to exactly suit the items produced, standardized containers and holding devices serve as a mediating element.

This is particularly true for the handling of "organic matter" (in Giedion's formulation). For the early period of the implementation of conveyor belts in the processing of cattle and hog carcasses, Giedion was able to see that the complete mechanization of the slaughtering, skinning, and dissection process was not possible at that time.

Just how far adaptation to standardized processing systems, with or without intermediate containers, has progressed in the intervening years is shown in impressive fashion in the documentary film *Our Daily Bread* by Austrian filmmaker and producer Nikolaus Geyrhalter. In quiet, concentrated images without any spoken commentary, the film shows how everything is adapted to box forms, made processable and stackable in the course of its transition from life to food, whether salad, chickens, fish, or pigs. Scientific breeding methods ensure a continually greater standardization of forms, weights, and sizes. Newer machines adapt to forms more flexibly. Only when it comes to batch processing does the container help.

Before the goods find their way to the end user, they normally disappear into their packaging. Consequently, a description of the container principle must include packaging. On the one hand, it forms the last link in the chain that binds the container system from production to the sale of goods. Its dimensions, derived from optimal stackability and compatibility with the volumes of larger transport units, are mostly the only evidence that the normal end user receives of the existence of the container system.

On the other hand, nothing has gotten members of affluent societies so acclimated to the rule of the container principle as the standardized packaging of everyday consumer goods that fill the shelves of supermarkets row after row with only slight variation on the same basic forms, and whose labels usually provide the only connection to the food purchased. This training effect is intensified by the raw sales culture of the low-cost supermarket. It forgoes shelves altogether and instead uses the next biggest components of the transport logistics system as sales surfaces: the cartons, boxes, and pallets on which the containers are transported into the store.

The development of containers into packaging (and from there into a brand) is mostly a matter of the relationship between objects and their labels. And containers can hardly be regarded as revolutionary subjects. Nevertheless, in view of the radical reversal of relations in the course of this history, it seems appropriate to speak of a revolution. The introduction of (industrially produced) packaging for the (industrially produced) bulk commodity means the emancipation of the container—a fundamental adjustment of the relationship between packaging and contents. It is closely associated with the rise of brands and the specific form of modern consumer capitalism.

Packaging serves as a link between the production and the sale of a product, and consequently it becomes a communications medium that informs the buyer about the good made invisible by the packaging and that exclusively sets the product apart from any other. (Thus, as in many other container configurations, the closedness of the container is a requirement for the utilization of a specific container function.) In his book *The Total Package: The Secret History and Hidden Meanings of Boxes, Bottles, Cans, and Other Persuasive Containers,* American journalist and author Thomas Hine considers the basis and cultural meanings of packaging. His depiction of their functions reminds us once again of black boxes:

The idyllic interaction of conveyor logic and the container principle is shown in the allegorical depiction of agricultural production on the frontispiece of a stock share certificate of the railway corporation Grand Union Company in 1968. The railway line is shown as an endless band on which the foodstuffs, depicted in their natural state on the other side of the image, glide into the cities' distribution centers, portioned, preserved, packaged, and canned in a procession of standardized cartons, cans, and bottles. Today, the organic matter often enters a container transport network in which everything is coordinated and mechanically enmeshed, from the plastic crate to the container. © Grand Union Company.

Packaging helps people know and decide things quickly, a task it accomplishes by a combination of display and concealment....

From this point of view, such instant recognition and understanding reveals the effective package to be an advanced technology and a model for the many complex information interfaces—from industrial control systems to five-hundred-channel cable menus—that are part of contemporary life.[20]

The packaging and its labels, with all their particularizing, identity-forming traits and applications—form, color, material, script, images, symbols, and so forth—are the site and manifestation of the brands, of the symbolic body of the product. In this way, the container acquires a dual nature. As packaging, it is the container of a product; it forms the physical barrier between the product and the outside world. However, the packaging is also the medium that physically carries the brand, which exists apart from the product contained within. The brand materialized in the packaging is the persona, the carrier of the social function of the goods. In this guise, it enters the company of the other goods and their buyers on the market.

The brand is a transcendental factor of consumer capitalism, and packaging is the material condition of its possibility. Nowadays one speaks of "communities" when people rally around symbolically powerful brands, or of "brand zones," established for the maintenance of a "rite of shopping," as Hiromi Hosoya and Markus Schaefer explain.[21]

What seemed to be a global dominance of brands (and not only in the eyes of capitalism's critics) is the divestment of capitalist value-added production "from the world of things," and the move toward the "conceptual value added" of products, as Naomi Klein explains in her book *No Logo: Taking Aim at the Brand Bullies*.[22] This overpowering representative function, which attaches itself to the neutral container and not to the specific product, causes us to buy brands and not products. The attraction of the names and images on the packaging is greater than that of the products.

Andy Warhol was one of the first to recognize this. Through his works of art composed of detergent boxes and soup cans, he reflected back to the modern consumer society a principle central to its creation of value. Shifting the container to the position of the item is tantamount to a revolution in the order of things to buy. The most fundamental criticism of the superficiality of modern consumer capitalism is sparked by this development. Conversely, however, it also forms the basis of the most affirmative practices of a consumer avant-garde, for whom shopping bags—another important form of the metacontainer—serve as a signal of membership in certain social groups.

What happens if the "use value" of a good (which, according to Karl Marx's famous definition, produces the "commodity character" of products in economic circulation, together with the "exchange value") is reduced (almost) completely to the display of precisely this consumption, as in the case of brand consumption?[23] What if the use value is targeted at the exchange value itself, expressed in money and in prestige or social status? The relation of exchange value to use value is radicalized once more. If one conceives of the commodity form as a container of the exchange value, then the manifestation of the brands strengthens this container function to the point at which it materializes on the packaging.

This is not to deny, of course, that several of the most prominent representatives of this brand metaphysics go without packaging altogether. In the case of clothing or sneakers, for instance, a few characters and names are enough to invoke the brand aura

In the container supermarket, the vision of a component of the global logistics network comes into its own. This is a magazine advertisement for the long-established shipping company American President Lines, now a daughter company of the Singapore-based shipper Neptune Orient Lines. © American President Lines. From *Containerisation International*, 40th anniversary issue, 2007.

and to radicalize the self-referentiality of the exchange value already laid out in Marx's analysis.

With mass-produced industrial packaging, with the standardized single-use container, the problem of garbage is exacerbated and becomes a further sign of consumer capitalism. For instance, if one drives through Africa, one can recognize the level of development—that is, a nation's degree of poverty—by whether the landscape is full of plastic bags. In the poorest nations, there is a lack of even basic containers, and as a result nothing is simply thrown away.

The chronology, places, and cultural conditions for the historical development into modern packaging can be determined fairly precisely, because (industrial and early industrial) packaging has always been gladly disposed of in the nearest countryside. Thus, in England, for example, the motherland of industrialization, riverbanks near cities are veritable archaeological sites for researching everyday life in early industrial society. Collecting historical bottles and cans has become a widespread hobby—and not only in England.

Bertolt Brecht wrote the following in American exile in 1945 during the last months of the war:

High in the clouds the feverish goods climb the mountain pass.
The tollgate is brittle, a thousand years old; they trample it
    down.
CHEAP is their password. The old men there! Do priests come to curse the wicked? No, they come to buy. The walls there! Never overtaken! With balls of light cotton shrewd agents blow the Chinese walls away
With a smile.[24]

This passage is part of an (incomplete) attempt to demonstrate the "unnatural character of bourgeois relations"[25] through a didactic poem in hexameter. The core of the text, as well as the quote at the beginning of the chapter, is formed by a rendering of *The Communist Manifesto* by Karl Marx and Friedrich Engels (1848).

Brecht assumes a strangely anachronistic position. In an ancient linguistic form, he reformulates what was then a 100-year-old classic text of modernity, and in so doing anticipates the central aspects of today's critique of brand consumption and globalization, 50 years later. The image of self-moving containers, of religions that surrender before the cult of consumption, of the smilingly bombed Chinese walls, is easily transferred to the present. Above all, however, the depiction of goods as agents with a certain subjectivity and capacity to act—already laid out by Marx—finds unexpected resonance, not the least because of the fact that with the fall of the Soviet Union, Marx and friends seemed to be entirely relegated to the sidelines.

In the future, even more efficient ordering and logistics systems could make it more affordable for the large retailers to close their stores and switch completely to the direct delivery of goods to customers. For now, however, development is going in a direction that strengthens the communicative aspect of packaging and is leading to a further subjectification of goods. "Intelligent" packaging (so-called smart packages) could become a central element of new, highly technologized self-service markets. When a shopper touches a product on the shopping cart's video screen, an advertisement would be played, and a digital shopping list could lead the shopper to the corresponding shelves by transmitting data via an RFID chip, where the goods would communicate their inventory and their prices, point out sales, and suggest why the purchase of an alternative product may be worthwhile. Or a package of medication could inform its buyer about individual dosing instructions.

Here, the container function of packaging as persona is united with the vision of the Internet of things (see chapter 6). In this vision of the container world of the future, containers communicate with containers—with packaging, support frames, storage containers, and displays. The products gain agency, albeit only to the extent to which their digital identity markers (i.e. product information) are detached from the material carriers (and their potential use values).

Container Subjects

In Aki Kaurismäki's film *The Man without a Past*, a man newly arrived in Helsinki, who has lost all of his belongings, his identity, and almost his life in a robbery, finds refuge in a container colony near the harbor. With modest means and with no knowledge of his past, he begins a new life among society's poor and outcast. A key scene is a moment in which we see him next to a newly planted garden in front of his container, when rain sets in as if by divine confirmation of his deed, watering the freshly planted bed of potatoes.

If modern man has become mad from his metaphysical unsettledness, as Sloterdijk writes, then a man who has lost all of his belongings and even his memory is the most extreme example of homelessness.[26] If he succeeds, as in the case of Kaurismäki's protagonist, in feeling at home again in some way, then he epitomizes victory over uprootedness or at least demonstrates the basic possibility of living in nothingness.

The container, as foundationless and mobile as it might be, it is the minimal embodiment of "four walls," thus giving symbolic form to the principal absence of a house. In this way, however, it also becomes the cornerstone for an attempt at a post-metaphysical, rationalized housing creation, as I show in chapter 7. Containers gather together the scattered things produced by an unfettered analytical operation that never ceases to dismember all areas of the living and the dead into individual pieces.

Civilization by container, as in the modular concepts of modern architecture or in the agglomeration of packaged goods in the modern household, stands diametrically opposed to the unified ideal that may still have been prevalent at the beginning of modernity, in the early nineteenth century. The container is a shell for the heterogeneous, for the explosive mixture—in microcosm, as a base unit for (mass) housing, and in macrocosm, as a spatial form for shaping the modern environment.

The container system combines the linear, one-dimensional logic of the digital with the two-dimensional logic of the chart and the spatial logic of packing and stacking. To use a term made popular by journalist Thomas Friedman's book *The World Is Flat*, it operates on the boundary between the "flat" and "not flat" world. As a principle of standardized spatial processing, it underlies both the tendency toward the total fluidity of pure administration in electronic data processing as well as the gravity and inertia, the finiteness and decline of the material world.

The mobilized, standardized container order, the "serial discipline of the box" that the artist Allan Sekula wrote about in "Fish Story", his large photo and text project on global economy and containerization, involves not only a spatial but also a temporal dimension.[27] The unsettling rolling of the ground that Brecht's poem cited at the beginning of this chapter has a fragmented nature—it happens in packaged, serial, mobile units. The supply chain is a chain of separate links. Spatial standardization and temporal sequencing yield the temporal-spatial mobilization of objects and beings specific to the container system.

Analogously, the lives of modern, modularized people are increasingly fashioned as a chain (or perhaps multiple parallel chains) of discrete links. American studies scholar John G. Blair precisely examined the roots of modularization in various areas of U.S. society in the nineteenth century, from industry to school and university education to music and the novel. He shows how this organizational principle came about and how it functions, dividing a complex process into flexible, standardized units, such that its parts can be exchanged and recombined with other compatible and equivalent pieces: "The whole is not fragmented; it is conceived in terms of subsystems or components."[28]

According to Blair, it is this organizational structure, and not what people are or what they believe, that led to a specifically "American way." Today, in the wake of significant EU unification projects like the Bologna Process and the Program for International Student Assessment Studies, the schools and universities in the old states of Europe are grappling with the question of the best form of modularized education. However, something bigger may be behind this: the reenactment of a development that has long since become a social reality in other areas—the modular biography organization of people in modern societies, called upon to regularly reinvent themselves and live their lives serially in stages. Should this actually be the case, it would have to be described as a legacy of the "American age" (by all appearances drawing to a close).

"Contain yourself" is the slogan of an American chain called The Container Store. The company, which sells nothing but packaging, exemplifies the emancipation of the container. Apart from this, the slogan expresses a subjectification strategy. The phrase invokes the notion of an autonomous individual.

In his study of the rise of the modern "disciplinary society," composed in the 1960s, French philosopher Michel Foucault argued that the isolation of individuals through cell-shaped spatial organization, temporal fragmentation, and a combinatorics of strength organized according to division of labor led to a maximal form of social control of the individual.

"The soul is the prison of the body," he wrote on the effect of the internalization programs shaped by pietism and Calvinism, which represented a fundamental requirement for this maximization of power effects in early modern society. The crisis subject of the "disenchanted" (Western) world at the beginning of the twenty-first century, emptied of all large social programs, can no longer be uplifted through one's soul, from the inside to the outside. In the age of the "flexible person" and modular lives, of temporally, spatially, and substantially mobilized biographies, the so-called interior no longer has enough enduring substance (and is also consequently no longer suited to be a stable disciplinary organ).[29]

The interior is regularly exchanged, in part or in whole. The establishment of the subject takes place from the outside inward, in accordance with a container model of filling with contents and encasing with additional protective, armorlike shells. (Strictly speaking, there is no longer this distinction between inner perception and outer perception, either; humans, whose subjectivity is defined by a container function, perceive themselves as if from the edge of a bowl and will take up the same kind of objectifying distance inward as outward.) The container-shaped constitutional form corresponds to a general buildup of gadgets of all kinds that enrich and control external (and internal) perception. This could be advanced electronic minidevices or functional clothing.

In principle, even the containers sold in the Container Store could be counted among the preservers of container subjectivity. In his dystopian vision of a "capsular society" of control and zonal spatial planning dominated by container subjectivities, Belgian architectural theorist and philosopher Lieven De Cauter writes:

> A society of mobility is inconceivable without omnipresent control. Whereas the disciplinary society was based on internalization, the control society functions externally, through militarization of urban space.... To an increasing degree, transport is becoming the transit between controlled and enclosed zones. The generic city is obsessed by closure, safety and control. One can appropriately term this the cellular city, and even the capsular civilization. The simultaneously archaic and hypermodern "archetypal face" of twenty-first century architecture and urbanism will be the enclosure, the wall, the barrier, the gate, the fence, the fortress....
>
> All means of transport beyond a certain level of speed ... become capsules: the train, the automobile, the aeroplane and, obviously, the space capsule. These are real capsules. Besides these there are also virtual capsules, such as a screen, a Walkman, a mobile phone. The omnipresence of screens (television screens, computer screens, ... windscreens) is part of capsularization. One could go as far as to say that each screen creates its own time-space milieu, whether virtual or not.[30]

The closed world of isolated and mutually regulating Cold War power blocs—in which every citizen of the world, at least in the Northern Hemisphere, defined him- or herself through affiliation with one of the two giant containers (analyzed by American computer and technological historian Paul Edwards for its mutual dependence on information models and politics)—becomes a world of partial territories and islands drifting apart and scattered individuals who are boxed in, encapsulated in their control and security containers.

Advanced consumer culture in nesting doll mode—the Container Store sells nothing but containers. Flyer of the New York branch of the chain on Broadway, 2006. © 2003 The Container Store Inc.

Thus, in the development from the containment of postwar policy to the global supply chains of today, under the banner of the collapse of the communist power sphere in the globalized world, there are two contrary tendencies: while the importance of nation-states conceived of as containers and of policies that are strictly tied to national territories have decreased sharply in favor of a multilevel network, there has been and continues to be a tremendous growth of microcontainers. It seems then, as though a system of comparatively few stationary, large, and medium-size "containers" (states, corporations, and institutions) has been replaced by a system with a tremendous multitude of smaller "containers" (mobile, flexible, more or less independent corporate units).

With the belief in the permanence of states and their institutions, bureaucracies and statelike corporations, the last bastion of collective sense making through large, quasi-transcendent units has fallen. But must we see the development into isolated container subjectivities so negatively? After all, the encapsulation as confinement outlined (as a cautionary tale) by De Cauter represents only one of many forms. The subject, cobbled together from technologies and (certified) proficiencies, strives to synthesize the most diverse and contradictory things: cosmopolitanism *and* local rootedness; manual skill *and* abstract intellectuality; artistic creativity *and* pragmatic realism; English, Chinese, the programming language C++, *and* ancient Greek. It does not simply live in multiple places, it also exists in multiple subjectivities that swap their respective settings. Subject containers, places and containers for distinct subjectivizations, conform to (or serve) the container subject—driver-self, triathlete-self, papa-self, computer worker–self.

The container subject is an assumption, a self-modeling, the attempt to assemble a (robust, or at least durable) subjectivity from component parts, since the classic, indefinite subjectifications (laid out for the entire life span) are not available. However the establishment of the self on the outside goes on the market, it fights for meager resources. The subject containers of modularized biographical organization are comparable to the boxes in Erving Goffman's sociological analysis of the "territories of the self"—exclusive claims of ownership attached to an object, like a chair or a table, or to a specific space, like a telephone booth or a tennis court. These egocentric reservations for the establishment and stabilization of the self must be continually marked; otherwise they will be commandeered by others.

In the meantime, the ancient cultural practices—from the first vessels for fire, water, and the remains of the deceased to the universal container for everything everywhere—have also given birth to cities, if we give credence to technological historian Lewis Mumford, and nations, if we follow the sociological critics of the container theory of society like Ulrich Beck.[31] The departure from the large-scale forms of the collective and the development of the container principle as a complex systemic network of partially autonomous, neighborly, and temporally limited groupings of related container individuals is accompanied by the hope that these loose and only temporary forms of subjectification may not lead to a further encapsulation.

For in the conceptual impossibility of the neologism *container individual* itself, the scandal that this view of consistently tentative living conditions represents is revealed, measured by the traditional notions of subjectivity and biographical unity, which continue to determine our values. *Individual* means "the undivided." Containers, by contrast, always consist of themselves as a shell and of their contents. Thus, they decompose into at least two parts, of which one part means nothing, apart from containing something (comparable to the fact of mere being, living simply by virtue of not having died), and the other part bears a plethora of meanings but cannot contain them for long.

In contrast, the *dividual*, the subject that owns up to its principle division and compounded nature, would perhaps no longer need the fiction (or the principle) of the container at all, if it established itself as a kind of consolidation in the networks and changing forms of the collective. As a transitional project between encapsulation and programmatic temporal and spatial division stands the container subject, "with sober eyes," isolated, bottomless, gathering, and holding itself together.

Retired containers in the service of the global nature-culture system: fortification of shifting sand dunes on the beach in Veracruz, Mexico, March 1994. Photograph from Allan Sekula, *Fish Story* (Düsseldorf, Germany: Richter Verlag, 2002), 159. Used with permission. © Allan Sekula.

Notes

Preface

1. Jim Abrams, "Law Requiring 100 Percent Cargo Screening Sets Tough Standards," Associated Press, August 22, 2007, Department of Homeland Security, http://www.homelandcouncil.org/news.php?newsid=1256.

2. U.S. Customs and Border Protection, *Container Security Initiative: 2006–2011 Strategic Plan*, http://www.cbp.gov/linkhandler/cgov/border_security/international_activities/csi/csi_strategic_plan.ctt/ csi_strategic_plan.pdf.

3. Susan Sontag, "The Imagination of Disaster," in *Against Interpretation and Other Essays* (New York: Farrar, Straus & Giroux, 1966).

Introduction

1. Bernhard Siegert, "Cacography or Communication? Cultural Techniques in German Media Studies," *Grey Room* 29 (Fall 2007): 30.

Chapter 1

1. Andy Warhol, *The Philosophy of Andy Warhol: From A to B and Back* (London: Penguin, 2007), 145.

2. John W. Smith, "Saving Time: the Archives of the Andy Warhol Museum," 1996. http://www.carnegiemuseums.org/cmag/bk_issue/1996/janfeb/warhol.html (accessed on July 3rd 2006).

3. G. Edward Pendray, "Archaeology for the Future Now Being Sealed in Crypts," *Science News Letter* (September 17, 1938), 179–180; see William E. Jarvis, *Time Capsules* (Jefferson, N.C.: McFarland, 2003), 85.

4. Robert Ascher, "How to Build a Time Capsule," *Journal of Popular Culture* 8, no. 2 (1974): 241–253.

5. Konrad Köstlin, "Das Mass aller Dinge" [The measure of all things], *Du*, no. 733 (February 2003): 42–45.

6. Richard Pollak, *The Colombo Bay* (New York: Simon & Schuster, 2004), 12.

7. Paul Virilio, *Negative Horizon: An Essay in Dromoscopy* (London: Continuum, 2006), 77.

8. Paul Virilio, *The Original Accident* (Cambridge: Polity, 2007), 10.

9. Paul Virilio, *Unknown Quantity* (London: Thames and Hudson, 2003), 5

10. Érik Orsenna, *A Portrait of the Gulf Stream: In Praise of Currents* (London: Haus, 2010).

11. Hans-Jörg Rheinberger, "Natur, NATUR" [Nature, NATURE], in *Iterationen* [Iterations] (Berlin: Merve Verlag, 2005), 46.

Chapter 2

1. George M. Adams, Robert. C. King, and G. Lloyd Wilson, "The Freight Container as a Contribution to Efficiency in Transportation," *Annals of the American Academy of Political and Social Science* 187, no. 1 (1936): 27.

2. Walter Meyercordt, *Container-Fibel* [Container primer] (Mainz, Germany: Krausskopf, 1974), 11.

3. Ibid.

4. Marc Levinson, *The Box: How the Shipping Container Made the World Smaller and the World Economy Bigger* (Princeton, NJ: Princeton University Press, 2006), 127.

5. Axel Doßmann, Jan Wenzel, and Kai Wenzel, *Architektur auf Zeit: Baracken, Pavillons, Container* [Temporal architecture: barracks, gazebos, containers] (Berlin: B-Books, 2006), 24.

6. Peter Sloterdijk, *Sphären. Plurale Sphärologie*, Bd. III, Schäume . [Spheres: plural spherology, Vol. III: Foams] (Frankfurt am Main: Suhrkamp, 2004), 498.

7. Albert Einstein, "Foreword," in *Concepts of Space*, ed. Max Jammer (Cambridge, MA: Harvard University Press, 1954), xiii.

8. Hannes Böhringer, "Der Container" [The container], in *Orgel und Container* [Organ and container] (Berlin: Merve Verlag, 1993), 20. Translation by Barbara Klose-Ullmann and Alexander Klose.

9. George Lakoff and Mark Johnson, *Metaphors We Live By* (Chicago: University of Chicago Press, 2003), 25 and 29.

10. Wilfred Bion, *Attention and Interpretation* (London: Maresfield Library, 1970); see Richard M. Billow, "Relational Levels of the 'Container-Contained' in Group Therapy," *Group* 24, no.4 (2000): 243–259.

11. etoy, *TANK-PLANT #5: Twisting Tangible Values*, 2005. http://etoy.TANK-PLANT #5.twisting tangible values.pdf.

12. Uwe Pörksen, *Weltmarkt der Bilder: Eine Philosophie der Visiotype* [World market of images: a philosophy of visiotypes] (Stuttgart, Germany: Klett-Cotta, 1997), 28.

13. "Container. Das Prinzip Globalisierung" [Container: the principle of globalization], *du* 733 (February 2003).

14. Martin Heidegger, "The Age of the World Picture," in *Off the Beaten Track*, ed. and trans. Julian Young and Kenneth Haynes (Cambridge, UK: Cambridge University Press, 2002).

15. Jakob Tanner, "Wirtschaftskurven. Zur Visualisierung des anonymen Marktes" [Economic curves: visualizing the anonymous market], in *Ganz normale Bilder. Historische Beiträge zur visuellen Herstellung von Selbstverständlichkeit* [Very normal images: historical contributions to the visual production of implicitness], ed. David Gugerli and Barbara Orland (Zurich: Chronos, 2002), 129–158.

16. Otto Neurath, "Bildliche Darstellung sozialer Tatbestände" [Pictorial representation of social facts], in *Bildersprache: Otto Neurath Visualisierungen* [Language of images: Otto Neurath's visualizations], ed. Erwin K. Bauer and Frank Hartmann (Vienna: Wiener Universitatsverlag, 2006), 6–11.

Chapter 3

1. Oliver E. Allen, "The Man Who Put Boxes on Ships," *Audacity: The Magazine of Business Experience*, Spring 1994, 15.

2. Ibid., 13.

3. Malcom P. McLean, "Opportunity Begins at Home," *American Magazine*, May 1950, 121.

4. *Quarterly Review* 63 (1839), 23; quoted by Wolfgang Schivelbusch, *The Railway Journey: The Industrialization of Time and Space in the 19th Century* (Berkeley: University of California Press, 1986), 10.

5. Bernhard Siegert, "Der Nomos des Meeres. Zur Imagination des Politischen und ihren Grenzen" [The Nomos of the sea: about the imagination of the political and its limits], in *Politiken der Medien* [Politics of media], ed. Daniel Gethmann and Markus Stauff (Zurich: Diaphanes), 40.

6. Carl Schmitt, *Land and Sea: A World-historical Reflection*, trans. Simona Draghici (Washington, DC: Plutarch, 1997), 51.

7. Ibid., 54.

8. Alfred Thayer Mahan, *Naval Strategy: Compared and Contrasted with the Principles and Practice of Military Operations on Land* (Boston: Little, Brown, 1911), 139.

9. Peter J. Hugill, *World Trade since 1431: Geography, Technology, and Capitalism* (Baltimore: Johns Hopkins University Press, 1993), 34 and 40.

10. Schmitt, *Land and Sea*, 9.

11. Friedrich Nietzsche, *The Gay Science: With a Prelude in German Rhymes and an Appendix of Songs* (Cambridge, UK: Cambridge University Press, 2001), 119.

12. Revelation 17:15 (AV).

13. Siegert, "Nomos," 48.

14. Francis Phillips, "The Containerising of America," *Containerisation International* (1981), 79.

15. Dieter Läpple, "The Port of Hamburg: Containerfloodgate or Logistic Service Centre?," in *Seaports in the Context of Globalization and Privatization*, ed. Rainer Dombois and Rainer Heiner Heseler (Bremen: University of Bremen, 2000), 87–104.

16. Peter Sloterdijk, *In the World Interior of Capital: Towards a Philosophical Theory of Globalization* (Cambridge, UK: Polity Press, 2013), 140.

17. See Arthur Donovan and Joseph Bonney, *The Box That Changed the World* (East Windsor, N.J.: Commonwealth Business Media, 2006), xxii.

18. Marc Levinson, *The Box: How the Shipping Container Made the World Smaller and the World Economy Bigger* (Princeton, N.J.: Princeton University Press, 2006), 53.

Chapter 4

1. Silvio Crespi, "Was ist ein Behälter?" [What is a container?], *Der Behälter*, January 1934.

2. Lewis Mumford, *The City in History* (New York: Harcourt, Brace, 1961), 14.

3. Ibid.

4. Peter Pachnicke, "Einführung" [Introduction], in *Welt der Gefässe von der Antike bis Picasso* [World of receptacles from antiquity to Picasso], ed. Bernhard Mensch and Peter Pachnicke (Oberhausen, Germany: Ludwig Galerie Schloss Oberhausen, 2004), 9.

5. Homer, *The Iliad*, trans. Richmond Lattimore (Chicago: University of Chicago Press, 2011), 24.527–533.

6. Hannes Böhringer, "Der Container" [The container], in *Orgel und Container* [Organ and container] (Berlin: Merve Verlag, 1993), 13.

7. Reinhold Bräuer and Max Krusemark, *Der Sprung aus dem Gleise: Der wirtschaftliche Kampf zwischen Auto und Reichsbahn—der Behälterverkehr* [Jumping the tracks: the economic battle between the freeway and railway—container transport] (Munich: Schweiszer, 1933), 16, 26, and 14.

8. Fritz Brauner, *Behälterverkehr* [Container transport] (Berlin: Elsner, 1933), 4–10, 12–17, 18.

9. Ibid., 4.

10. Konrad Köstlin, "Das Mass aller Dinge" [The measure of all things], *du*, February 2003, p. 45.

11. Claus Pias, "Wer sein Leben im Griff hat, kann einpacken: Der Umzugskarton als Medium der Selbstinventur" [If you master your life you can pack: the packing case as a medium of self inventory]. *Frankfurter Allgemeine Zeitung*, May 5, 1999.

Chapter 5

1. Thomas L. Friedman, *The World Is Flat: A Brief History of the Twenty-First Century* (New York: Farrar, Straus and Giroux, 2005), 151f.

2. Fred Scharmen, "Logistics: The Backend of the Big Box." Seven Six Five, 2006, http://www.sevensixfive.net/logistics/logistics.html.

3. Frederick Winslow Taylor, *The Principles of Scientific Management* (New York: Harper, 1913), 8.

4. Detlev Peukert, "Der 'Traum der Vernunft'" [The "dream of reason"], in *Max Webers Diagnose der Moderne* [Max Weber's diagnosis of modernity] (Göttingen, Germany: Vandenhoeck & Ruprecht, 1989), 74ff.

5. Reinhart Koselleck, "Die Geschichte der Begriffe und Begriffe der Geschichte" [The history of concepts and concepts of history], in *Begriffs-*

geschichten: Studien zur Semantik und Pragmatik der politischen und sozialen Sprache [Histories of concepts: studies on the semantics and pragmatics of political and social discourse] (Frankfurt am Main: Suhrkamp, 2006), 68f.

6. Rupert Scholtissek, "Rationalisieren" [Rationalizing], *Rationeller Transport* 11, no. 3/4 (1962): 6f.

7. Martin van Creveld, *Supplying War. Logistics from Wallenstein to Patton* (Cambridge, UK: Cambridge University Press, 2004), 1.

8. Dieter Läpple, "Vom Gütertransport zur logistischen Kette—Neue Anforderungen an Güterverkehrsnetze in einer international arbeitsteiligen Gesellschaft" [From cargo transport to chain logistics: new requirements for cargo transport networks in a society of international division of labor], *Mitteilungen der Deutschen Akademie für Städtebau und Landesplanung* 34, no. 1 (1990): 26.

9. Bernhard Siegert, *Relays: Literature as an Epoch of the Postal System*, trans. Kevin Repp (Redwood City, CA: Stanford University Press, 1999), 144, 154.

10. Ibid.

11. Protocol of the constitutional meeting of the Bureau International des Containers, Dec. 12th.1933, Paris; quoted in *Rationeller Transport*, Offizielles Organ der Studiengesellschaft für den kombinierten Verkehr e.V., 12. Jg., 1/1963, "Vor 30 Jahren" [30 years ago] [editorial], 1.

12. Dr. Meister, "Transportketten, integrierter Verkehr und Container—neue Schlüsselpunkte in der Wirtschaftslogistik" [Transport chains, integrated transit and containers: key points in business logistics], *Rationeller Transport* 18, no. 6 (1969): 267.

13. Michel Serres, *The Parasite*, trans. Lawrence R. Schehr (Minneapolis, MN: University of Minnesota Press, 2007), 21f.

14. Helmut Braun, Robert Obermaier, and Felix Müller, "Der Container als Artefakt eines Transportparadigmas: Akteure und Diffusionsphasen" [The container as a transport paradigm artifact: actors and diffusion phases], in *Logistikmanagement: Analyse, Bewertung und Gestaltung logis-*

tischer Systeme [Logistics management: analysis, evaluation and design of logistical systems], ed. Robert Obermaier and Andreas Otto (Wiesbaden, Germany: Deutscher Universitats-Verlag, 2007), 316.

15. Serres, *The Parasite*, 39.

16. Friedrich Kittler,"The History of Communication Media," *CTheory*, July 30, 1996, http://www.ctheory.net/articles.aspx?id=45.

17. Ibid.

18. Ulrich Welke, "Ladepläne und Fahrpläne" [Stowplanes and timetables], in *Seefahrt im Zeichen der Globalisierung* [Sea transport under the sign of globalization], ed. Heide Gerstenberger and Ulrich Welke (Münster, Germany: Westfälisches Dampfboot, 2002), 142.

Chapter 6

1. Wolfgang Pircher, "Krieg und Management: Zur Geschichte des Operations Research" [War and management: the history of operations research], in *Governmentality Studies: Analysen liberal-demokratischer Gesellschaften im Anschluss an Michel Foucault* [Governmentality studies: analyses of liberal-democratic societies in the wake of Foucault], ed. Ramón Reichert (Münster, Germany: Lit, 2004), 113.

2. Oskar Morgenstern, "Note on the Formulation of the Theory of Logistics," *Naval Research Logistics Quarterly* 2, no. 3 (1955): 133.

3. Edna Bonacich and Jake B. Wilson, *Getting the Goods: Ports, Labor, and the Logistics Revolution* (Ithaca, NY: Cornell University Press, 2008), 143f.

4. Joshua S. Gans, *Inside the Black Box: A Look at the Container* (Sydney, Australia: University of New South Wales Department of Economics, 1995), 12.

5. Foster L. Weldon, "Cargo Containerization in the West Coast–Hawaiian Trade," in *Operations Research* 6, no. 5 (September–October 1958): 649f.

6. W. Ross Ashby, "The Black Box," in *Introduction to Cybernetics* (New York: Wiley, 1956), 133.

7. James W. Cortada, *The Digital Hand: How Computers Changed the Work of American Manufacturing, Transportation, and Retail Industries* (New York: Oxford University Press, 2004), 228.

8. Frank Broeze, *The Globalisation of the Oceans* (St. Johns, Newfoundland: International Maritime Economic History Association, 2002), 23f.

9. Wolfgang Bohle, "Der kombinierte Verkehr in der betrieblichen Organisation einer Reederei" [The internal organization of bimodal traffic in a shipping company], *Containers*, no. 46 (March 1972).

10. Bernhard Siegert, *Passage des Digitalen. Zeichenpraktiken der neuzeitlichen Wissenschaften 1500–1900* [Passage of the digital: sign practices of the modern sciences, 1500–1900] (Berlin: Brinkmann und Bose, 1993), 43.

11. Steven Graham, "Excavating the Material Geographies of Cybercities," in *The Cybercities Reader* (London: Routledge, 2004), 139.

12. Christoph Neubert, "Onto-Logistik: Kommunikation und Steuerung im Internet der Dinge" [Onto-logistics: communication and control in the Internet of things], In *Agenten und Agenturen* [Agents and agencies], ed. Lorenz Engell, Bernhard Siegert, and Joseph Vogl (Weimar, Germany: Verlag der Bauhaus-Universität Weimar, 2008), 132.

13. Neal Stephenson, *Snow Crash* (New York: Bantam Dell, 2003), 20.

14. Ibid., 387

Chapter 7

1. Charles Jencks, *The New Paradigm in Architecture: The Language of Post-Modernism* (New Haven, CT: Yale University Press, 2002), 9.

2. Frank Lloyd Wright, "The Destruction of the Box: The Freedom of Space," in *Writings and Buildings* (New York: Horizon, 1966), 284f.

3. Ibid.

4. Bruno Taut, *Die Auflösung der Städte oder der Weg zur alpinen Architektur* [The dissolution of cities or: towards an alpine architecture] (Hagen, Germany: Folkwang-Verlag, 1920), n.p.

5. "Frank Lloyd Wright Talks on His Art," *New York Time Magazine*, October 4, 1953, 47; quoted in Richard Maccormac, "Form and Philosophy: Froebel's Kindergarten Training and Wright's Early Work," in *On and By Frank Lloyd Wright: A Primer of Architectural Principles*, ed. Robert McCarter (London: Phaidon Press, 2005), 129.

6. Walter Gropius, "Normung und Wohnungsnot" [Standardization and housing shortage], *Technik und Wirtschaft* 20, no. 1 (1927): 332.

7. Walter Rathenau, *Zur Kritik der Zeit* [On the critique of our time] (Berlin: Fischer, 1912), 15f.

8. Peter Sloterdijk, "Insulierungen" [Insulations], in *Sphären* [Spheres] *III. Schäume* [Foams] (Frankfurt am Main: Suhrkamp, 2004), 570.

9. Ibid., 504f.

10. Mies van der Rohe, 1924, quoted in Anne Hoormann, "Geometrische Module und kristalline Strukturen. Über den Irrtum einer Utopie aus dem Baukasten" [Geometrical modules and crystalline structures: on the fallacy of a construction kit utopia] in *Constructing Utopia. Konstruktionen künstlischer Welten* [Constructions of artificial worlds], ed Annett Zinsmeister (Zurich: Diaphanes, 2005), 51.

11. Walter Prigge, "Typologie und Norm: Zum modernen Traum der industriellen Fertigung von Wohnungen" [Typology and standard: on the modern dream of industrially manufactured housing], in *Constructing Utopia. Konstruktionen künstlicher Welten* [Constructions of artificial worlds], ed. Annett Zinsmeister (Zurich: Diaphanes, 2005), 74f.

12. Le Corbusier, *Feststellungen zu Architektur und Städtebau* (Braunschweig/Wiesbaden: Vieweg Verlag 1987), 91 and 94ff.

13. Ibid., 89

14. Ernst Neufert, *Architects' Data* (Chichester, UK: Wiley-Blackwell, 2012).

15. Ernst Neufert, *Bauordnungslehre* [Building regulations] (Berlin: Volk und Reich Verlag, 1943), n.p.

16. Hajo Eickhoff, "Haus" [House], in *Vom Menschen: Handbuch Historische Anthropologie* [On humankind: handbook of historical anthropology], ed. Christoph Wulff (Weinheim, Germany: Beltz, 1997), 221.

17. Ibid., 229.

18. Thomas P. Hughes, *American Genesis: A Century of Invention and Technological Enthusiasm, 1870–1970* (Chicago: University of Chicago Press, 2004), 254ff.

19. Winfried Nerdinger, *Der Architekt Walter Gropius* [The architect Walter Gropius] (Berlin: Mann, 1996), 22f.

20. Tom Wolfe, *From Bauhaus to Our House* (New York: Farrar, Straus and Giroux, 1981), 41ff.

21. Ibid., 9.

22. Ibid., 54.

23. Words and music by Malvina Reynolds. Copyright 1962 Schroder Music Co. (ASCAP). Renewed 1990. Used by permission. All rights reserved.

24. Keller Easterling, "Interchange and Container: The New Orgman," *Perspecta* 30 (1999): 113.

25. Colin Davies, *The Prefabricated Home* (London: Reaktion, 2005), 44.

26. Aaron Betsky, "From High-Tech to Lot-Ek: A Brief Journey," in *Lot-Ek: Mobile Dwelling Unit*, ed. Christopher Scoates (New York: Distributed Art Publishers, 2003), 94f.

27. Dieter Hoffmann-Axthelm, "Container, Behälter des Neuen" [Container, vessel of the new], in *Zeitzeichen Baustelle: Realität, Inszenierung und Metaphorik eines abseitigen Ortes* [Construction site as sign of the times: reality, staging and metaphorics of a remote space], ed. Franz Pröfener (Frankfurt am Main: Campus, 1998), 270f.

Chapter 8

1. Manuel Castells, *The Information Age: Economy, Society, and Culture* vol. 3, *End of Millennium* (Oxford, UK: Blackwell, 1998), 1.

2. Martin Heidegger, "The Thing," in *Poetry, Language, Thought* (New York: HarperCollins, 2001), 164.

3. Giorgio Agamben, *Homo Sacer: Sovereign Power and Bare Life*, trans. Daniel Heller-Roazen (Stanford, CA: Stanford University Press, 1998), 174.

4. Roger Willemsen, *Hier spricht Guantánamo* [Guantanamo speaking] (Frankfurt am Main: Fischer-Taschenbuch-Verlag, 2006), 11.

5. Axel Doßmann, Jan Wenzel, and Kai Wenzel, *Architektur auf Zeit : Baracken, Pavillons, Container* [Temporary architecture: baracks, gazebos, containers] (Berlin: B-Books, 2006), 24.

6. Tom Holert and Mark Terkessidis, "Erstarrte Mobilität" [Frozen mobility] (email interview by Axel Doßmann and Kai Wenzel), in *Architektur auf Zeit : Baracken, Pavillons, Container*, ed. Axel Doßmann, Jan Wenzel, and Kai Wenzel, Cologne Art Society (Berlin: B-Books, 2006), 67.

7. Manfred Studer, dir. *Klopfzeichen aus dem Container—blinde Passagiere auf Irrfahrt* [Tapping in the container—stowaways on an odyssey], SWR-TV, Baden-Baden, Germany, 2004.

8. Ibid.

9. Matei Bejenaru, *Travel Guide,* trans. Alex Moldovan (Berlin: Feinkost Galerie Berlin, 2008), n.p.

10. Roberto Saviano, *Gomorrah: A Personal Journey into the Violent International Empire of Naples' Organized Crime System* (New York: Picador, 2008), 3.

11. Ibid.

12. Peter Sloterdijk, *In the World Interior of Capital: Towards a Philosophical Theory of Globalization*, trans. Wieland Hoban (Cambridge, UK: Polity Press, 2013), xx.

13. Hannes Böhringer, "Der Container" [The container], in *Orgel und Container* [Organ and container] (Berlin: Merve Verlag, 1993), 19.

14. Niklas Luhmann, *Social Systems*, trans. John Bednarz Jr. with Dirk Baecker (Stanford, CA: Stanford University Press, 1995), 109.

15. Gaston Bachelard, *The Poetics of Space*, trans. Maria Jolas (Boston: Beacon Press, 1994), 88.

16. Sloterdijk, *In the World Interior of Capital*.

17. Quoted in Eugen Leitherer, "Warenverpackungen unter technischen und sozio-ökonomischen Kriterien" [Technical and socioeconomical criteria of goods packaging], in *Reiz und Hülle: Gestaltete Warenverpackungen des 19 und 20 Jahrhunderts* [Allurement and cover: designed goods packaging of the nineteenth and twentieth centuries], ed. Eugen Leitherer and Hans Wichmann (Basel, Switzerland: Birkhauser, 1987), 25f.

18. Siegfried Giedion, *Mechanization Takes Command* (New York: W. W. Norton, 1969), 3.

19. Ibid., 217.

20. Thomas Hine, *The Total Package: The Secret History and Hidden Meanings of Boxes, Bottles, Cans, and Other Persuasive Containers* (Boston: Little, Brown, 1995), 17f.

21. Hiromi Hosoya and Markus Schaefer, "Brand Zone," in *Harvard Design School Guide to Shopping*, ed. Chuihua J. Chung, Jeffrey Inaba, Rem Koolhaas, Sze Tsung Leong (Cologne, Germany: Taschen, 2001), 166.

22. Naomi Klein, *No Logo: Taking Aim at the Brand Bullies* (New York: Picador, 2000), 34.

23. Karl Marx, *Capital,* vol. 1, *A Critique of Political Economy*, trans. Ben Fowkes (New York: Penguin, 1992), 125.

24. Bertolt Brecht, "Lehrgedicht/ Das Manifest" [Didactic poem / the manifesto], in *Werke, Grosse kommentierte Berliner und Frankfurter Ausgabe* [Works, Great Annotated Berlin and Frankfurt edition], ed. Werner Hecht, Jan Knopf, Werner Mittenzwei, and Klaus-Detlef Müller, vol. 15, *Gedichte 5—Gedichte und Gedichtfragmente 1940–1956* [Poems 5: poems and poem fragments, 1940–1956] (Berlin: Suhrkamp, 1993), 153.

25. Bertolt Brecht, letter to Karl Korsch, April 1945, quoted in Manfred Wekwerth, "Manifest-Geschichten," *Ossietzky: Zweiwochenschrift für Politik / Kultur / Wirtschaft* (July 2006).

26. Sloterdijk, *In the World Interior of Capital*.

27. Allan Sekula, *Fish Story* (Düsseldorf, Germany: Richter, 2002), 55.

28. John G. Blair, *Modular America: Cross-Cultural Perspectives on the Emergence of an American Way* (Westport, CT: Greenwood Press, 1988), 3.

29. Michel Foucault, *Discipline and Punish: The Birth of the Prison* (New York: Vintage Books, 1995), 30.

30. Lieven De Cauter, "The Capsular Civilization: The City in the Age of Transcendental Capitalism," in *The Capsular Civilization: On the City in the Age of Fear* (Brussels, Belgium: NAi, 2004), 45f.

31. Ulrich Beck, *What Is Globalization?* (Cambridge, UK: Polity Press, 2000), 23.

Bibliography

Abrams, Jim. "Law Requiring 100 Percent Cargo Screening Sets Tough Standards." Associated Press, August 22, 2007. Department of Homeland Security, http://www.homelandcouncil.org/news.php?newsid=1256.

Adams, George M., Robert. C. King, and G. Lloyd Wilson. "The Freight Container as a Contribution to Efficiency in Transportation." *Annals of the American Academy of Political and Social Science* 187, no. 1 (1936): 27–36.

Agamben, Giorgio. *Homo Sacer: Sovereign Power and Bare Life.* Translated by Daniel Heller-Roazen. Stanford, CA: Stanford University Press, 1998.

Albrecht, Thorsten. *Truhen, Kisten, Laden—vom Mittelalter bis zur Gegenwart am Beispiel der Lüneburger Heide* [Chests, boxes, coffers: from the middle ages to the present using the example of Lüneburg Heath]. Petersberg, Germany: Michael Imhof, 1987.

Allen, Oliver E. "The Man Who Put Boxes on Ships." *Audacity: The Magazine of Business Experience* 2 (Spring 1994): 13–23.

Almond, Darren. *Index*. London: Koenig Books, 2008.

Altekamp, Stefan, and Knut Ebeling, eds. *Die Aktualität des Archäologischen in Wissenschaft, Medien und Künsten* [The topicality of the archeological in science, media and arts]. Frankfurt am Main: Fischer Verlag, 2004.

Arnold, Horace Lucien, and Fay Leone Faurote. *Ford Methods and the Ford Shops*. New York: Engineering Magazine Company, 1919.

Ascher, Robert. "How to Build a Time Capsule." *Journal of Popular Culture* 8, no. 2 (1974): 241–253.

Ashby, W. Ross. *Introduction to Cybernetics*. New York: Wiley, 1956.

Augé, Marc. *Non-Places: An Introduction to Supermodernity*. Translated by John Howe. London: Verso, 2009.

Bachelard, Gaston. *The Poetics of Space*. Translated by Maria Jolas. Boston: Beacon Press, 1994.

Bartz, Dietmar. "Bade-Entchens letztes Ufer" ["Rubber Ducks' Last Shore"]. *Neue Zürcher Zeitung*, April 2002.

Beck, Ulrich. *What Is Globalization?* Translated by Patrick Camiller. Cambridge, UK: Polity Press, 2000.

Behrendt, Siegfried, ed. *Computer, Internet und Co: Geld sparen und Klima schützen* [Computer, Internet and co.: how to save money and protect the climate]. Dessau-Rosslau, Germany: Umweltbundesamt, 2009.

Bejenaru, Matei. *Travel Guide*. Translated by Alex Moldovan. Berlin: Feinkost Galerie Berlin, 2008.

Beplat, Klaus H. *Megatrends in Containerisation: Directions and Projections*. Hamburg: Reinecke & Associates, 1989.

Bergdoll, Barry, and Peter Christensen, eds. *Home Delivery: Fabricating the Modern Dwelling*. New York: Museum of Modern Art, 2008.

Bernard, Andreas. *Lifted: A Cultural History of the Elevator*. New York: NYU Press, 2014.

Berz, Peter. "Die Wabe" [The honeycomb]. In *FAKtisch: Festschrift für Friedrich Kittler zum 60; Geburtstag* [FAKtual: Festschrift for Friedrich Kittler on the occasion of his 60th birthday], edited by Peter Berz, Annette Bitsch, and Bernhard Siegert, 65–81. Munich: Fink, 2003.

Betsky, Aaron. "From High-Tech to Lot-Ek: A Brief Journey." In *Lot-Ek: Mobile Dwelling Unit*, edited by Christopher Scoates, 87–98. New York: Distributed Art Publishers, 2003.

Billow, Richard M. "Relational Levels of the 'Container-Contained' in Group Therapy." *Group* 24, no. 4 (2000): 243–259.

Bion, Wilfred. *Attention and Interpretation*. London: Maresfield Library, 1970.

Blair, John G. *Modular America: Cross-Cultural Perspectives on the Emergence of an American Way*. Westport, CT: Greenwood Press, 1988.

Bloech, Jürgen, and Gösta B. Ihde, eds. *Vahlens Grosses Logistiklexikon* [Logistics dictionary] (Munich: Vahlen, 1997).

Blumenberg, Hans. "Lebenswelt und Technisierung unter Aspekten der Phänomenologie" [Environment and mechanization under phenomenological aspects]. In *Wirklichkeiten, in denen wir leben* [Realities we live in], 7–54. Stuttgart, Germany: Reclam, 1999.

Bohle, Wolfgang. "Der kombinierte Verkehr in der betrieblichen Organisation einer Reederei" [The internal organization of bimodal traffic in a shipping company]. *Containers*, no. 46 (March 1972): 45–57.

Böhringer, Hannes. "Der Container" [The container]. In *Orgel und Container* [Organ and container], 7–34. Berlin: Merve Verlag, 1993.

Bolz, Norbert, Friedrich Kittler, and Christoph Tholen, ed. *Computer als Medium* [Computer as medium]. Munich: Wilhelm Fink, 1994.

Bonacich, Edna, and Jake B. Wilson. *Getting the Goods: Ports, Labor, and the Logistics Revolution*. Ithaca, NY: Cornell University Press, 2008.

Bonney, Joseph, and Arthur Donovan. *The Box That Changed theWorld*. East Windsor, NJ: Commonwealth Business Media, 2006.

Bräuer, Reinhold, and Max Krusemark. *Der Sprung aus dem Gleise: Der wirtschaftliche Kampf zwischen Auto und Reichsbahn—der Behälterverkehr* [Jumping the tracks: The economic battle between the freeway and railway—container transport]. Munich: Schweiszer, 1933.

Braun, Helmut, Robert Obermaier, and Felix Müller. "Der Container als Artefakt eines Transportparadigmas: Akteure und Diffusionsphasen" [The container as a transport paradigm artifact: actors and diffusion phases]. In *Logistikmanagement: Analyse, Bewertung und Gestaltung logistischer*

Systeme [Logistics management: analysis, evaluation and design of logistical systems], edited by Robert Obermaier and Andreas Otto, 309–345. Wiesbaden, Germany: Deutscher Universitats-Verlag, 2007.

Brauner, Fritz. *Behälterverkehr* [Container transport]. Berlin: Elsner, 1933.

Brecht, Bertolt. "Lehrgedicht"/ "Das Manifest" [Didactic poem / the manifesto]. In *Werke, Grosse kommentierte Berliner und Frankfurter Ausgabe* [Works, great commented Berlin and Frankfurt edition]. Edited by Werner Hecht, Jan Knopf, Werner Mittenzwei, Klaus-Detlef Müller. Vol. 15, *Gedichte 5—Gedichte und Gedichtfragmente 1940–1956* [Poems 5: poems and poem fragments 1940–1956], , 152–153. Berlin: Suhrkamp, 1993.

Broeze, Frank. *The Globalisation of the Oceans*. St. Johns, Newfoundland: International Maritime Economic History Association, 2002.

Brunn, Stanley D., ed. *Wal-Mart World: The World's Biggest Corporation in the Global Economy*. New York: Routledge, 2006.

Bührer, K. W. *Raumnot und Weltformat* [Shortage of space and world format]. Munich: Die Brücke, 1912.

Carr, Nicholas. *The Big Switch: Rewiring the World, from Edison to Google*. New York: W. W. Norton, 2008.

Castells, Manuel. *The Information Age: Economy, Society, and Culture*, Volume 3, *End of Millennium* Oxford, UK: Blackwell, 1998.

Castells, Manuel. "Space of Flows, Space of Places: Materials for a Theory of Urbanism in the Information Age." In *The Cybercities Reader*, edited by Stephen Graham, 82–93. London and New York: Routledge, 2004.

Certeau, Michel de. *The Practice of Everyday Life*. Translated by Steven Rendall. Berkeley: University of California Press, 1984.

Ceruzzi, Paul E. *A History of Modern Computing*. 2nd ed. Cambridge, MA: MIT Press, 2003.

Clark, James L. "Refrigeration Snags Reduced to 3 Phases." *Container News*, April 1973.

Conrad, Joseph. "Ocean Travel." In *Last Essays*, 53–58. London: J. M. Dent & Sons, 1926.

"Container Ships." *Marine Engineering and Shipping Age*, October 1932.

Cortada, James W. *The Digital Hand: How Computers Changed the Work of American Manufacturing, Transportation, and Retail Industries.* New York: Oxford University Press, 2004.

Crespi, Silvio."Was ist ein Behälter?" [What is a container?]. *Der Behälter*, January 1934.

Creveld, Martin Van. *Supplying War: Logistics from Wallenstein to Patton.* Cambridge, UK: Cambridge University Press, 2004.

Culemeyer, Johann. *Die Eisenbahn ins Haus: Die Beförderung von Eisenbahnwagen und Schwerlasten mit Strassenfahrzeugen* [The train to the house: transport of rail cars and heavy loads by road vehicles], edited by Alfred Gottwaldt. Düsseldorf, Germany: VDI Verlag, 1987.

Davies, Colin. *The Prefabricated Home.* London: Reaktion, 2005.

De Cauter, Lieven. *The Capsular Civilization: On the City in the Age of Fear.* Brussels, Belgium: NAi, 2004.

Deecke, Helmut. "Expressdienste als Vorreiter der Industrialisierung des Gütertransports" [Express services as trailblazers for the industrialization of cargo transport]. In *Güterverkehr, Logistik und Umwelt: Analysen und Konzepte zum interregionalen und städtischen Verkehr* [Cargo transport, logistics and environment: analyses and concepts on interregional and urban traffic], edited by Dieter Läpple, 59–83. Berlin: Edition Sigma, 1993.

Derrida, Jacques. "No Apocalypse, Not Now (Full Speed Ahead, Seven Missiles, Seven Missives)." Translated by Philip Lewis and Catherine Porter. *Diacritics* 14, no. 2, (Summer 1984): 20–31.

Derrida, Jacques. *The Post Card: From Socrates to Freud and Beyond.* Chicago: University of Chicago Press, 1987.

Donovan, Arthur, and Joseph Bonney. *The Box That Changed the World.* East Windsor, N.J.: Commonwealth Business Media, 2006.

Donovan, Arthur, and Andrew Gibson. *The Abandoned Ocean: A History of United States Maritime Policy*. Columbia, SC: University of South Carolina Press, 2000.

Doßmann, Axel, Jan Wenzel, and Kai Wenzel. *Architektur auf Zeit: Baracken, Pavillons, Container* [Temporal architecture: barracks, gazebos, containers]. Berlin: B-Books, 2006.

Easterling, Keller. "Interchange and Container: The New Orgman." *Perspecta* 30 (1999): 112–121.

Ebbesmeyer, Curtis C., and W. James Ingraham, Jr. "Pacific Toy Spill Fuels Ocean Current Pathway Research." *Eos, Transactions, American Geophysical Union* 75, no. 37 (September 7, 1994): 425–431.

Ebbesmeyer, Curtis C., and W. James Ingraham, Jr. "Shoe Spill in the North Pacific." *Eos, Transactions, American Geophysical Union* 73, no. 34 (August 25, 1992): 361–365.

Edwards, Paul N. *The Closed World: Computers and the Politics of Discourse in Cold War America*. Cambridge, MA: MIT Press, 1996.

Edwards, Paul N. "From 'Impact' to Social Process: Computers in Society and Culture." In *Handbook of Science and Technology Studies*, edited by Sheila Jasanoff, 257–285. Thousand Oaks, CA: Sage, 1995.

Eickhoff, Hajo. "Haus" [House]. In *Vom Menschen: Handbuch Historische Anthropologie* [On humankind: handbook of historical anthropology], edited by Christoph Wulff, 221–230. Weinheim, Germany: Beltz, 1997.

Eilers, Reimer. *Das neue Tor zur Welt: Vierzig Jahre Container im Hamburger Hafen* [The new gate to the world: 40 years of containers in the port of Hamburg]. Hamburg: Mare Verlag, 2009.

Einstein, Albert. "Foreword." In *Concepts of Space*, edited by Max Jammer, xiii–xvii. Cambridge, MA: Harvard University Press, 1954.

etoy. "Mission Eternity: Crossing the Deadline." http://www.missioneternity.org.

etoy. *Tank Plant #5: Twisting Tangible Values*, 2005. http://etoy.TANKPLANT #5.twisting tangible values.pdf.

Fiedler, Jeanine, ed. *Social Utopias of the Twenties: Bauhaus, Kibbutz and the Dream of the New Man*. Wuppertal, Germany: Muller and Busmann Press, 1995.

Foucault, Michel. *Discipline and Punish: The Birth of the Prison*. New York: Vintage Books, 1995.

Friedman, Thomas L. *The World Is Flat: A Brief History of the Twenty-First Century*. New York: Farrar, Straus and Giroux, 2005.

Galhena, Ravinda. "Intermodal Revolution." *Containerisation International*, 40th anniversary issue (2007): 41–45.

Gans, Joshua S. *Inside the Black Box: A Look at the Container*. Sydney, Australia: University of New South Wales Department of Economics, 1995.

Gerstenberger, Heide, and Ulrich Welke, eds. *Seefahrt im Zeichen der Globalisierung* [Sea transport under the sign of globalization]. Münster, Germany: Westfälisches Dampfboot, 2002.

Geyrhalter, Nikolaus, dir. *Unser Täglich Brot* [Our daily bread]. Vienna: Nikolaus Geyrhalter Filmproduktion, 2005.

Giedion, Siegfried. *Mechanization Takes Command*. New York: W. W. Norton, 1969.

Goffman, Erving. *The Presentation of Self in Everyday Life*. Garden City, NY: Doubleday, 1959.

Graham, Steven. "Excavating the Material Geographies of Cybercities." In *The Cybercities Reader*, 138–42. London: Routledge, 2004.

Grobecker, Kurt. *800 Jahre Hafen Hamburg: Das offizielle Jubiläumsbuch* [800 years of Hamburg harbor: the official anniversary book]. Hamburg: Christians, 1988.

Gropius, Walter. "Normung und Wohnungsnot" [Standardization and housing shortage]. *Technik und Wirtschaft* 20, no. 1 (1927).

Gugerli, David. "Soziotechnische Evidenzen. Der 'pictorial turn' als Chance für die Geschichtswissenschaft" [Sociotechnical evidences: The 'pictorial turn' as a chance for historical science]. In *Traverse* 3/1999, "Wissenschaft, die Bilder schafft" [Science that produces images], 131–159.

Gugerli, David. *Suchmaschinen. Die Welt als Datenbank* [Search engines: The world as a database]. Frankfurt: Suhrkamp, 2009.

Hägermann, Dieter, and Helmuth Schneider. *Propyläen Technikgeschichte, Erster Band: Landbau und Handwerk 750 v. Chr.–1000 n. Chr* [Propyläen history of technology, first volume: agriculture and crafts 750 B.C.–1000 A.D.]. Berlin: Propyläen-Verlag, 1997.

Hapag-Lloyd, ed. *Kleine Geschichte der Seepost von den Anfängen bis 1914* [A short history of maritime mail from the beginnings to 1914]. Hamburg: Hapag-Lloyd, 1986.

Heidegger, Martin. "The Age of the World Picture." In *Off the Beaten Track*, edited and translated by Julian Young and Kenneth Haynes, 57–85. Cambridge, UK: Cambridge University Press, 2002.

Heidegger, Martin. "The Thing." In *Poetry, Language, Thought*, 161–184. New York: HarperCollins, 2001.

Hellmann, Ullrich. *Künstliche Kälte: Die Geschichte der Kühlung im Haushalt* [Artificial cold: the story of domestic cooling]. Giessen, Germany: Anabas-Verlag, 1990.

Henry, James J., and Henry J. Karsch. "Container Ships." In *Containerisation: A Modern Transport System*, edited by G. Van den Burg. London: Hutchinson, 1969.

Hine, Thomas. *The Total Package: The Secret History and Hidden Meanings of Boxes, Bottles, Cans, and Other Persuasive Containers*. Boston: Little, Brown, 1995.

Hoffmann-Axthelm, Dieter. "Container, Behälter des Neuen" [Container, vessel of the new]. In *Zeitzeichen Baustelle: Realität, Inszenierung und Metaphorik eines abseitigen Ortes* [Construction site as sign of the time: reality, staging and metaphorics of a remote space], edited by Franz Pröfener, 266–71. Frankfurt am Main: Campus, 1998.

Holert, Tom, and Mark Terkessidis. "Erstarrte Mobilität" [Frozen mobility] (email interview by Axel Doßmann and Kai Wenzel). In *Architektur auf Zeit : Baracken, Pavillons, Container*, ed. Axel Doßmann, Jan Wenzel, and Kai Wenzel, Cologne Art Society. Berlin: B-Books, 2006.

Holert, Tom, and Mark Terkessidis. "Was bedeutet Mobilität?" [What does mobility mean?]. In *Projekt Migration* [Project migration], edited by Kölnischer Kunstverein, 98–107. Cologne, Germany: DuMont, 2005.

Homer. *The Iliad*. Translated by Richmond Lattimore. Chicago: University of Chicago Press, 2011.

Hompel, Michael ten. "'Das Internet der Dinge': Kartons und Kisten werden zu eigenständigen Objekten — A' ["The Internet of things": cartons and boxes become autonomous objects using RFID technology]. *Deutsche Verkehrs Zeitung*, May 31, 2005, 28–29.

Hoormann, Anne. "Geometrische Module und kristalline Strukturen. Über den Irrtum einer Utopie aus dem Baukasten" [Geometrical modules and crystalline structures: On the fallacy of a construction kit utopia]. In *Constructing Utopia. Konstruktionen künstlischer Welten* [Constructions of artificial worlds], edited by Annett Zinsmeister, 51–53. Zurich: Diaphanes, 2005.

Hoppe, Ralf. "Der globale Legostein: Der Container hat die Weltwirtschaft revolutioniert—weil er so schön praktisch ist" [The global Lego brick: The container revolutionized global economy—because it is so convenient]. *Kultur Spiegel* 11 (2001): 20–22.

Hosoya, Hiromi, and Markus Schaefer. "Brand Zone." In *Harvard Design School Guide to Shopping*, edited by Chuihua J. Chung, Jeffrey Inaba, Rem Koolhaas, Sze Tsung Leong,164–73. Cologne, Germany: Taschen, 2001.

Hughes, Thomas P. *American Genesis: A Century of Invention and Technological Enthusiasm, 1870–1970*. Chicago: University of Chicago Press, 2004.

Hugill, Peter J. *World Trade since 1431: Geography, Technology, and Capitalism*. Baltimore: Johns Hopkins University Press, 1993.

Ignarski, Sam, ed. *The Box: An Anthology Celebrating 25 Years of Containerisation and the TT Club*. London: EMAP Business Communications, 1995.

Jack in the Box, Container Architecture. http://www.koelnerbox.de/architektur.

Jarvis, William E. *Time Capsules: A Cultural History*. Jefferson, NC: McFarland, 2003.

Jencks, Charles. *The New Paradigm in Architecture: The Language of Post-Modernism*. New Haven, CT: Yale University Press, 2002.

Kaissling, Karl. "1928–1968: 40 Jahre Forschungsarbeit für den kombinierten Verkehr" [1928–1968: 40 years of research for bimodal traffic]. *Rationeller Transport* 17, no. 5 (1968): 427–30.

Kaurismäki, Aki, dir. *Der Mann ohne Vergangenheit* [The man without a past]. Helsinki, Finland: Sputnik, 2002.

Kittler, Friedrich. "Die Stadt ist ein Medium." In *Mythos Metropole*, edited by Gotthard Fuchs, Bernhard Moltmann, and Walter Prigge, 228–244. Frankfurt: Suhrkamp, 1995.

Kittler, Friedrich. "The History of Communication Media." *CTheory*, July 30, 1996, http://www.ctheory.net/articles.aspx?id=45.

Klaus, Peter. *Die dritte Bedeutung der Logistik* [The third meaning of logistics]. Hamburg: Deutscher Verkehrs-Verlag, 2002.

Klein, Naomi. *No Logo: Taking Aim at the Brand Bullies*. New York: Picador, 2000.

Kneissl, Peter. "Die utriclarii: Ihre Rolle im gallo-römischen Transportwesen und Weinhandel" [The utriclarii: Their role in gallo-roman cargo transport and wine-trade]. *Bonner Jahrbücher* 181 (1981): 169–203.

Koselleck, Reinhard. "Die Geschichte der Begriffe und Begriffe der Geschichte" [The history of concepts and concepts of history]. In *Begriffsgeschichten: Studien zur Semantik und Pragmatik der politischen und sozialen Sprache* [Histories of concepts: Studies on the semantics and pragmatics of political and social discourse], 56–76. Frankfurt am Main: Suhrkamp, 2006.

Köstlin, Konrad. "Das Mass aller Dinge" [The measure of all things]. *du*, February 2003.

Kuhn, Gerd. "Die Spur der Steine: Über die Normierung des Ziegelsteins, das Oktametersystem und den 'Maszstab Mensch'" [Tracking the stones: about the standardization of the brick, the oktameter system and the "human scale"]. In *Ernst Neufert: Normierte Baukultur im 20 Jahrhundert* [Ernst Neufert: standardized building culture in the twentieth century], edited by Walter Prigge, 334–357. Frankfurt am Main: Campus, 1999.

Küster, Hansjörg. *Geschichte des Waldes: Von der Urzeit bis zur Gegenwart* [History of the woods: From the primeval to the present]. Munich: C. H. Beck, 2003.

Lakoff, George, and Mark Johnson. *Metaphors We Live By*. Chicago: University of Chicago Press, 2003.

Läpple, Dieter. "Essay über den Raum: Für ein gesellschaftswissenschaftliches Raumkonzept" [Essay on space: In favor of a social scientific concept of space]. In *Stadt und Raum* [City and space], edited by H. Häussermann, 157–208. Pfaffenweiler, Germany: Centaurus-Verlagsgesellschaft, 1991.

Läpple, Dieter. "The Port of Hamburg: Containerfloodgate or Logistic Service Centre?" In Seaports in the Context of Globalization and Privatization, edited by Rainer Dombois and Rainer Heiner Heseler, 87–104. Bremen: University of Bremen, 2000.

Läpple, Dieter. "Vom Gütertransport zur logistischen Kette—Neue Anforderungen an Güterverkehrsnetze in einer international arbeitsteiligen Gesellschaft" [From cargo transport to chain logistics: new requirements for cargo transport networks in a society of international division of labor]. *Mitteilungen der Deutschen Akademie für Städtebau und Landesplanung* 34, no. 1 (1990): 11–33.

Le Corbusier. *Feststellungen zu Architektur und Städtebau* [Statements on architecture and urban development]. Braunschweig/Wiesbaden: Vieweg Verlag, 1987.

Le Corbusier. *The Ideas of Le Corbusier on Architecture and Urban Planning*. Translated by Margaret Guiton. New York: George Braziller, 1981.

Le Corbusier. *Toward an Architecture*. Translated by John Goodman. Los Angeles: Getty Research Institute, 2007.

Le Corbusier, *Vers une architecture*. Paris: Éditions Vincent Fréal & Cie, 1966.

Leitherer, Eugen. "Geschichte der Markierung und des Markenwesens" [History of branding]. In *Die Marke: Symbolkraft eines Zeichensystems* [The brand: symbolic power of a sign system], edited by Manfred Bruhn, 55–74. Bern, Switzerland: Haupt, 2001.

Leitherer, Eugen. "Warenverpackungen unter technischen und sozio-ökonomischen Kriterien" [Technical and socioeconomical criteria of goods packaging]. In *Reiz und Hülle: Gestaltete Warenverpackungen des 19 und 20 Jahrhunderts* [Allurement and cover: designed goods packaging of the nineteenth and twentieth centuries], edited by Eugen Leitherer and Hans Wichmann, 9–111. Basel, Switzerland: Birkhauser, 1987.

Leong, Sze Tsung. "Mobilize." In *Harvard Design School Guide to Shopping*, edited by Chuiha J. Chung, Jeffrey Inaba, Rem Koolhaas, Sze Tsung Leong, 500–503. Cologne, Germany: Taschen, 2001.

Levinson, Marc. *The Box: How the Shipping Container Made the World Smaller and the World Economy Bigger*. Princeton, NJ: Princeton University Press, 2006.

Lindner, Erik. *Die Herren der Container: Deutschlands Reeder-Elite* [The lords of the containers: Germany's shipowner elite]. Hamburg: Hoffmann und Campe, 2008.

Luhmann, Niklas. *Social Systems*. Translated by John Bednarz Jr. with Dirk Baecker. Stanford, CA: Stanford University Press, 1995.

Lutz, Hans-Rudolf. *Die Hieroglyphen von heute: Grafik auf Verpackungen für den Transport* [The hieroglyphs of today: graphics on transport packaging. Zurich: Hans-Rudolf Lutz Verlag, 1990.

Maccormac, Richard. "Form and Philosophy: Froebel's Kindergarten Training and Wright's Early Work." In *On and by Frank Lloyd Wright: A Primer of Architectural Principles*, edited by Robert McCarter, 124–143. London: Phaidon Press, 2005.

Mahan, Alfred Thayer. *Naval Strategy: Compared and Contrasted with the Principles and Practice of Military Operations on Land*. Boston: Little, Brown, 1911.

Marx, Karl. *Capital*. Vol. 1, *A Critique of Political Economy*. Translated by Ben Fowkes. New York: Penguin, 1992.

Mathes, Heinz Dieter. "Rationalisierung" [Rationalization]. In *Handwörterbuch der Wirtschaftswissenschaft* [Concise dictionary of economics]. Vol. 6. Edited by Willi Albers, 399–406. Stuttgart, Germany: Fischer Verlag, 1981.

McLean, Malcom. "Opportunity Begins at Home." *American*, May 1950.

Meister, Dr. "Transportketten, integrierter Verkehr und Container—neue Schlüsselpunkte in der Wirtschaftslogistik" [Transport chains, integrated transit and containers: key points in business logistics]. *Rationeller Transport* 18, no. 6 (1969): 267.

Melville, Herman. *The Confidence-Man: His Masquerade*. New York: W. W. Norton, 2006.

Mensch, Bernhard, and Peter Pachnicke, eds. *Welt der Gefässe von der Antike bis Picasso* [World of vessels from antiquity to Picasso]. Oberhausen, Germany: Ludwig Galerie Schloss Oberhausen, 2004.

Meyercordt, Walter. *Behälter und Paletten: Flurfördermittel, Lager- und Betriebseinrichtungen* [Containers and pallets: loading devices, storage and enterprise facilities]. 2nd ed. Darmstadt, Germany: Hestra, 1964.

Meyercordt, Walter. *Container-Fibel* [Container primer]. Mainz, Germany: Krausskopf, 1974.

Miler, Zdenek, and J. A. Novotny. *Krtek va meste* [The mole in the city] [Cartoon film]. Prague, Czechoslovakia: Krátký Film Praha, 1982.

Miller, Rich. "The Great Debate: Data Center Containers." Data Center Knowledge, January 6, 2009, http://www.datacenterknowledge.com/archives/2009/01/06/the-great-debate-data-center-containers.

Morgenstern, Oskar. "Note on the Formulation of the Theory of Logistics." *Naval Research Logistics Quarterly* 2, no. 3 (1955): 129–136.

Mumford, Lewis. *The City in History*. New York: Harcourt, Brace, 1961.

National Museum of American History. Bruce Seaton in conversation with Arthur Donovan. Containerization Oral History Collection, 1995–1998, Smithsonian Institution, Washington, DC.

Nerdinger, Winfried. *Der Architekt Walter Gropius* [The architect Walter Gropius]. Berlin: Mann, 1996.

Neubert, Christoph. "Onto-Logistik: Kommunikation und Steuerung im Internet der Dinge" [Onto-logistics: communication and control in the Internet of things]. In *Agenten und Agenturen* [Agents and agencies], edited by Lorenz Engell, Bernhard Siegert, and Joseph Vogl, 119–33. Weimar, Germany: Verlag der Bauhaus-Universität Weimar, 2008.

Neufert, Ernst. *Architects' Data*. Chichester, UK: Wiley-Blackwell, 2012.

Neurath, Otto. "Bildliche Darstellung sozialer Tatbestände" [Pictorial Representation of Social Facts]. In *Bildersprache: Otto Neurath Visualisierungen* [Language of images. Otto Neurath's visualizations], edited by Erwin K. Bauer and Frank Hartmann, 6–11. Vienna: Wiener Universitatsverlag, 2006.

Nietzsche, Friedrich. *The Gay Science: With a Prelude in German Rhymes and an Appendix of Songs*. Cambridge, UK: Cambridge University Press, 2001.

Orsenna, Érik. *A Portrait of the Gulf Stream: In Praise of Currents*. London: Haus, 2010.

Ortiz, David S., Susan E. Martonosi, and Henry H. Willis. "Evaluating the Viability of 100 Percent Container Inspection at America's Ports." In *The Economic Impacts of Terrorist Attacks*, edited by Peter Gordon, James E. Moore II, and Harry W. Richardson, 218–241. Cheltenham, UK: Elgar, 2005.

Panofsky, Dora, and Erwin Panofsky. *Pandora's Box: The Changing Aspects of a Mythical Symbol*. New York: Pantheon Books, 1956.

Peacock, D. P. S., and D. F. Williams. *Amphorae and the Roman Economy*. New York: Longman, 1986.

Peukert, Detlev. "Der 'Traum der Vernunft'" [The "dream of reason"]. In *Max Webers Diagnose der Moderne* [Max Weber's diagnosis of modernity], 55–91. Göttingen, Germany: Vandenhoeck & Ruprecht, 1989.

Phillips, Francis. "The Containerising of America." In *The Box: An Anthology Celebrating 25 Years of Containerisation and the TT Club*, edited by Sam Ignarski, 75–79.

Pias, Claus. "Wer sein Leben im Griff hat, kann einpacken: Der Umzugskarton als Medium der Selbstinventur" [If you master your life you can pack The moving box as a medium of self inventory]. *Frankfurter Allgemeine Zeitung*, May 5, 1999.

Pircher, Wolfgang. "Krieg und Management: Zur Geschichte des Operations Research" [War and Management: On the history of operations research]. In *Governmentality Studies: Analysen liberal-demokratischer Gesellschaften im Anschluss an Michel Foucault* [Governmentality studies: dissections of liberal-democratic societies following Michel Foucault], edited by Ramón Reichert, 113–25. Münster, Germany: Lit, 2004.

Poet, Paul, dir. *Ausländer Raus! Schlingensief's Container* [Foreigners out! Schlingensief's container]. [Documentary film]. Austria, 2001.

Pölking-Eiken, Hermann-J., Hartmut Schwerdtfeger, and Thomas von Zabern. *Bremen/Bremerhaven Container Story: Die Erfolgsgeschichte einer Kiste, die den Hafen veränderte* [Bremen/Bremerhaven Container Story: Success story of a box that changed the port], edited by the Bremer Lagerhaus-Gesellschaft. Bremen, Germany: Steintor, 1991.

Pollak, Richard. *The Colombo Bay*. New York: Simon & Schuster, 2004.

Pörksen, Uwe. *Weltmarkt der Bilder: Eine Philosophie der Visiotype* [World market of images: a philosophy of visiotypes]. Stuttgart, Germany: Klett-Cotta, 1997.

Preuss, Olaf. *Eine Kiste erobert die Welt* [A box conquers the world]. Hamburg: Murmann, 2007.

Prigge, Walter, ed. *Ernst Neufert. Normierte Baukultur im 20. Jahrhundert* [Ernst Neufert: Standardized building culture in the twentieth century], Frankfurt/M u. New York: Campus, 1999.

Prigge, Walter. "Typologie und Norm: Zum modernen Traum der industriellen Fertigung von Wohnungen" [Typology and standard: on the modern dream of industrially manufactured housing]. In *Constructing Utopia: Konstruktionen künstlicher Welten* [Constructions of artificial worlds], edited by Annett Zinsmeister, 69–77. Zurich: Diaphanes, 2005.

Rathenau, Walter. *Zur Kritik der Zeit* [On the critique of our time]. Berlin: Fischer, 1912.

Reynolds, Malvina. "Little Boxes." *Little Boxes and Other Handmade Songs*. Oak Publishing, 1962.

Rheinberger, Hans-Jörg. "Natur, NATUR" [Nature, NATURE]. In *Iterationen* [Iterations], 30–50. Berlin: Merve Verlag, 2005.

Rushton, Alan, Phil Croucher, and Peter Baker. *The Handbook of Logistics and Distribution Management*, 5th ed. London: Kogan Page, 2014.

Samwel, Emad. "IT: Past, Present and Future." *Containerisation International*, 40th anniversary issue (2007): 51–57.

Samwel, Emad. "Taking Its Time: The Slowness of Implementation of RFID Projects Is Leading Many in the Shipping Industry to Believe That RFID Has Failed." *Containerisation International* (May 2007): 74.

Saviano, Roberto. *Gomorrah: A Personal Journey into the Violent International Empire of Naples' Organized Crime System*. New York: Picador, 2008.

Schabacher, Gabriele. "Raum-Zeit-Regime: Logistikgeschichte als Wissenszirkulation zwischen Medien, Verkehr und Ökonomie" [Space-time-regimes: logistics history as knowledge circulation between media, traffic and economy]. In *Agenten und Agenturen* [Agents and agencies], edited by Lorenz Engell, Bernhard Siegert, and Joseph Vogl, 135–48. Weimar, Germany: Verlag der Bauhaus-Universität Weimar, 2008.

Scharmen, Fred. "Logistics: The Backend of the Big Box." Seven Six Five, 2006, http://www.sevensixfive.net/logistics/logistics.html.

Schivelbusch, Wolfgang. *The Railway Journey: The Industrialization of Time and Space in the 19th Century*. Berkeley: University of California Press, 1986.

Schmidt-Sommerfeld, Georg-Wilhelm. "Vierzig Jahre Internationales Behälterbüro aus deutscher Sicht" [40 years of the International Container Office from a German perspective]. *Die Bundesbahn* 2 (1973): 97–104.

Schmitt, Carl. *Land und Meer: Eine weltgeschichtliche Betrachtung* [Land and sea: a world-historical reflection]. Stuttgart, Germany: Klett-Cotta, 1993.

Scholtissek, Rupert. "Rationalisieren" [Rationalizing]. *Rationeller Transport* 11, no. 3/4 (1962): 6f.

Schwedt, Georg. *Vom Tante-Emma-Laden zum Supermarkt: Eine Kulturgeschichte des Einkaufens* [From mom-and-pop store to supermarket: A cultural history of shopping]. Weinheim, Germany: Wiley-VCH, 2006.

Sekula, Allan. *Fish Story*. Düsseldorf, Germany: Richter, 2002.

Serres, Michel. *The Parasite*. Translated by Lawrence R. Schehr. Minneapolis: University of Minnesota Press, 2007.

Siegert, Bernhard. "Cacography or Communication? Cultural Techniques in German Media Studies." *Grey Room* 29 (Fall 2007): 26–47.

Siegert, Bernhard. "Cultural Techniques: Or the End of the Intellectual Postwar Era in German Media Theory." *Theory, Culture & Society* 30, no. 6 (2013): 48–65.

Siegert, Bernhard. "Der Nomos des Meeres. Zur Imagination des Politischen und ihren Grenzen" [The Nomos of the sea: about the imagination of the political and its limits]. In *Politiken der Medien* [*Politics of Media*], edited by Daniel Gethmann and Markus Stauff, 39–56. Zurich: Diaphanes.

Siegert, Bernhard. *Passage des Digitalen: Zeichenpraktiken der neuzeitlichen Wissenschaften, 1500–1900* [Passage of the digital. Sign practices of the modern sciences 1500–1900]. Berlin: Brinkmann & Bose, 2003.

Siegert, Bernhard. *Relays: Literature as an Epoch of the Postal System*. Trans. Kevin Repp. Redwood City, CA: Stanford University Press, 1999.

Sloterdijk, Peter. *In the World Interior of Capital: Towards a Philosophical Theory of Globalization*. Translated by Wieland Hoban. Cambridge, UK: Polity Press, 2013.

Sloterdijk, Peter. "Schäume" [Foams]. In *Sphären III* [Spheres III]. Frankfurt am Main: Suhrkamp, 2004.

Smith, John W. "Andy Warhol." In *Deep Storage: Arsenale der Erinnerung—Sammeln, Speichern, Archivieren in der Kunst* [Deep Storage: arsenals of memory: collecting, storing, archiving in arts], edited by Ingrid Schaffner and Matthias Winzen, 278–281. Munich: Prestel, 1997. (For a shorter English version, see Smith, John W. "Saving Time: The Archives of the Andy Warhol Museum." Carnegie Museums, January-February 1996, http://www.carnegiemuseums.org/cmag/bk_issue/1996/janfeb/warhol.html.)

Sontag, Susan. "The Imagination of Disaster." In *Against Interpretation and Other Essays*. New York: Farrar, Straus & Giroux, 1966.

"Special Kühlgüter," *Port of Hamburg*, February 2005.

Stephenson, Neal. *Snow Crash*. New York: Bantam Dell, 2003.

Studer, Manfred, dir. *Klopfzeichen aus dem Container—blinde Passagiere auf Irrfahrt* [Tapping in the container: stowaways on an odyssey], SWR-TV, Baden-Baden, Germany,.2004.

Sun Microsystems. *Sun Modular Datacenter: The World's First Virtualized Datacenter*. 2007. http://de.sun.com/sunnews/events/2007/blackbox-tour/index.jsp

Taggart, Stewart. "The 20-Ton Packet: Ocean Shipping Is the Biggest Real-Time Data-Streaming Network in the World." *Wired*, October 1999.

Tanner, Jakob. "Wirtschaftskurven: Zur Visualisierung des anonymen Marktes" [Economics curves: visualizing the anonymous market]. In *Ganz normale Bilder: Historische Beiträge zur visuellen Herstellung von Selbstverständlichkeit* [Very normal images: historical contributions to the visual production of implicitness], edited by David Gugerli, David Orland, and Barbara Orland, 129–158. Zurich: Chronos, 2002.

Tati, Jacques, dir. *Playtime*. Paris: Jolly Film, 1967.

Taut, Bruno. *Die Auflösung der Städte oder der Weg zur alpinen Architektur* [The dissolution of the cities or the way to alpine architecture]. Hagen, Germany: Folkwang-Verlag, 1920.

Taylor, Frederick Winslow. *The Principles of Scientific Management*. New York: Harper, 1913.

Thorby, Chris. "Freight Forwarders and Logistics: Evolution and Revolution." *Containerisation International*, 40th anniversary issue (2007): 21–27.

Tschoeke, Jutta. "Frostige Glieder: Aspekte der Kühlkette" [Frosty limbs: aspects of the cold chain]. In *Unter Null: Kunsteis, Kälte und Kultur* [Subzero: Artificial ice, cold and culture], edited by Center for Industrial Culture Nuremberg and Munich City Museum, 129–141. Munich: C. H. Beck, 1991.

U.S. Customs and Border Protection. *Container Security Initiative: 2006–2011 Strategic Plan*. Available at http://www.cbp.gov/linkhandler/cgov/border_security/international_activities/csi/csi_strategic_plan.ctt/ csi_strategic_plan.pdf (accessed on December 8, 2008).

Virilio, Paul. "The Accident Museum." In *A Landscape of Events*, translated by Julie Rose, 54–61. Cambridge, MA: MIT Press, 2000.

Virilio, Paul. *Negative Horizon: An Essay in Dromoscopy*. London: Continuum, 2008.

Virilio, Paul. *The Original Accident*. Cambridge: Polity, 2007.

Virilio, Paul. *Unknown Quantity*. London: Thames & Hudson, 2003.

Warhol, Andy. *The Philosophy of Andy Warhol: From A to B and Back*. London: Penguin, 2007.

Weigend, Michael. "Intuitive Modelle der Informatik" [Intuitive models of informatics]. PhD dissertation, University of Potsdam, 2007, http://www.informatikdidaktik.de/examensarbeiten/Weigend2007.pdf.

Weldon, Foster L. "Cargo Containerization in the West Coast–Hawaiian Trade." *Operations Research* 6, no. 5 (September-October 1958): 649–670.

Welke, Ulrich. "Ladepläne und Fahrpläne" [Loading plans and timetables]. In *Seefahrt im Zeichen der Globalisierung* [Sea transport under the sign of globalization], edited by Heide Gerstenberger and Ulrich Welke, 136–145. Münster, Germany: Westfälisches Dampfboot, 2002.

Wiborg, Klaus, and Susanne Wiborg. *1847–1997: Unser Feld ist die Welt; 150 Jahre Hapag-Lloyd* [1847–1997: Our field is the world; 150 years of Hapag-Lloyd]. Hamburg: Hapag-Lloyd, 1997.

Willemsen, Roger. *Hier spricht Guantánamo* [Guantánamo speaking]. Frankfurt am Main: Fischer-Taschenbuch-Verlag, 2006.

Winterbottom, Michael, dir. *In this World*. London: BBC, 2002. Ipoque, 2007, http://www.ipoque.de/userfiles/file/p2p_study_2007_abstract_de.pdf.

Witthöft, Jürgen. *Container: Eine Kiste macht Revolution* [Container: A box stages revolution]. Hamburg: Koehler, 2000.

Wolfe, Tom. *From Bauhaus to Our House*. New York: Farrar, Straus and Giroux, 1981.

Wright, Frank Lloyd. "The Destruction of the Box: The Freedom of Space." In *Writings and Buildings*, 284–289. New York: Horizon, 1966.

Zinsmeister, Annett, ed. *Constructing Utopia. Konstruktionen künstlischer Welten* [Constructions of artificial worlds]. Zurich: Diaphanes, 2005.

Zweig, Philip L. *Wriston: Walter Wriston, Citibank, and the Rise and Fall of American Financial Supremacy*. New York: Crown, 1995.

Index

Abrams, Jim, vii
Accidents. *See also* Shipwrecked
 containers
 Hansa Carrier, 29–30
 MSC Napoli, 1–6
 relationship to technology,
  26–29, 34–35
Advertisements
 American President Lines, 330
 The Container Store, 335–336
 German Federal Railway
  (*Deutsche Bahn AG*), 121
 global transport companies,
  297–299
 LIFT-VANS, 118–119
 mail-order houses, 285–286
 pallets, 187
 Project Blackbox, 199, 202
 Sea-Land Services, Inc., 96
 Seatrain, 104–105
 standardized housing, 285
 Wake-Up Wal-Mart, 318
Agamben, Giorgio, 305
Albert I (King of Monaco), 32
al-Harith, Jamal, 306
"All red system," 94
Almond, Darren, 294–296

Amazon, 242
American Challenge, 50
American-European relationship
 in architecture, 273–282
 container transport, 49–55, 109–
  110, 186–188, 321
American homeland security, vii,
 318–322
American Idol TV show, 305
American Institute of Architects,
 254, 255
Americanism, 54, 273–274
American President Lines (APL),
 106–108, 330
American transport. *See also* Sea-
 Land Services, Inc.
 compared to European, 49–55,
  109–110, 186–188, 321
 east and west coast, 104
 economic model, 98–99
"American way," 335
Am Horn development plans,
 259–261
Amphorae, 129–138
 distinguished from other containers, 129–131, 137
 Globular Amphora Culture, 127

history of, 133–135
markings, 132–133, 135
for olive oil, 134–135
transition to barrels, 135–138
types of, 132, 137–138
Andy Warhol Museum, 11
Animal slaughter, 324–325
Annals of the American Academy of Political and Social Science, 45
Anthropotope, 62
Apocalypse metaphor, 101–103, 108–109
Applications for containers, 59–63
Der Arbeiter, 81
Archaeology
  shipwrecks as resource, 35–36
  time capsules as resource, 16–18
Archigram architectural group, 283, 289
Architects
  Gropius, Walter, 259–262, 271, 274, 275
  Le Corbusier, 64, 248–250, 263–264, 266, 271–272, 289
  Mies van der Rohe, Ludwig, 263–264
  Wright, Frank Lloyd, 255–258
Architects' Data, 265
Architecture
  American hegemony, 273–282
  box architecture, 157–159, 253, 255–258, 278–282
  cell-based, 64, 249–251, 258–270, 274, 276, 281–282
  classical, 256–257, 263, 270
  container-based, 59–60, 287–292, 305, 309, 333–334
  high-rise, 251–255
  international style, 273, 275–277

inter-war period, 270–273
isolated, 253–255
modern (first wave), 255–270
modern (second-wave), 251–253, 283–292
modular, 250, 261–262, 265, 271, 275, 282–284
residential preferences, 251–255, 278–280, 284–292, 333–334
Argonauts, 105
Armour Institute of Chicago, 275
Around-the-world transport services, 113–117. *See also* Globalization
Art
  containers in, 70–74, 259–261, 278–279, 294–296, 301–304, 314–315, 329
  philosophy of, 66, 129, 138, 308
The Art of War (Précis de l'art de la guerre), 169–170
Ascher, Robert, 15–16
Ashby, Ross, 216, 219
Asian shipping, 18, 104, 110, 114
Asylum and refugee seekers. *See* Refugees and asylum seekers
Athens Charter, 274–275
Augé, Marc, 277
Automation. *See* Fordism

Baader, Ritter von, 142
Bachelard, Gaston, 316
Barrels
  construction of, 137
  distinguished from amphorae, 129–131, 137
  history of, 135–137
  terminology, 139
Barrier, Albert, 180

"Basic Concepts in History," 165
Bauhaus Dessau foundation, 263, 271
 Weimar, 271
Bauhaus school of design, 259, 264
 and Americanism, 275
Bauordnungslehre (Building Regulation), 265–267
Beck, Ulrich, 340
Bee metaphor, 261
Behälter, 55–56
 compared to American container, 50–54
 history of terminology, 139–141
Der Behälter (The Container) publication, 28–29
Behrens, Peter, 249–250
Bejenaru, Matei, 312–313
Bell-Beaker Culture, 127
Benjamin, Walter, 76
Benning, James. *See RR* documentary film
Betsky, Aaron, 287
Beuys, Joseph, 71
BIC *(Bureau International des Containers)*, 39, 41–42, 274–275
 definition of container transport, 42–47, 178
Big Brother TV show, 302, 304
The Big Switch: Rewiring the World, from Edison to Google, 242
Bion, Wilfred, 67
Bit-Torrent, 245
Black box model, 69, 234, 326–328
 in communications, 316–317
 cybernetic principles, 216–219

Project Blackbox (Sun Microsystems), 199–200, 202, 228, 229, 240, 242
Blair, John G., 335
Blumenberg, Hans, 234
Bohle, Wolfgang, 224–225
Böhringer, Hannes, 66, 138, 315
Bologna Process, 335
Box architecture, 251–255
 destruction of, 255–258
Boxes
 compared to containers, 150–152
 definition of, 147–148
 empty-container logistics, 195
 folding instructions, 152
 history of, 146–149
Box landscapes, 76–79
Box space, 65
Box stores, 157
Brand consumption, 328–332
Bräuer, Reinhold, 142–145
Braun, Helmut, 182
Brauner, Fritz, 147–148
Brecht, Bertolt, 331–332
Bremen, Germany, 47–49, 125, 224
British Empire, 94–95, 97–98
Broeze, Frank, 221
Broken transit, 181–184
Die Brücke organization, 268–269
Brunel, Isambard Kingdom, 92
Building Regulation (Bauordnungslehre), 265–267
Bureau International des Containers (BIC), 39, 41–42, 274–275
 definition of container transport, 42–47, 178
Bush, George W., 306

Camp X-Ray, 306
Canary Islands, 308
Capitalism, 4, 34, 84
 and packaging, 326–331
Capital Lease, 196
Capsular society, 337
Cargonauts, 104–105
Carr, Nicholas, 242–243
Cassa mobile, 40–42
Cast Away (film), 21–22
Castells, Manuel, 297
Ceramic Neolithic Age, 127
Chains, logistical, 176–180, 335
Chests, history of, 147–152
Chinese Postman Problem, 160
CIMC group, 196
Classification system for intermodal containers, 147–148. *See also* Coding of containers
Cloud computing, 240–247
Coding of containers, 29, 42
 Roman amphorae, 131, 133, 135
Cold chain *(chaîne de froid),* 179–180
Cold War period, 204, 304, 321, 337
Combined-transport containers. *See* Intermodal containers
The Communist Manifesto, 332
Compagnie Générale Maritime (CGM), 28
Computer. *See* Data processing
Computer-assisted design (CAD) programs, 268
Computer-Oriented Method for Planning and Process Control in the Seaport (COMPASS), 224

Confidence Man (Herman Melville), 101–102
Conrad, Joseph, 93
Consumer capitalism, 4, 34
 and packaging, 326–331
Container. *See* Container ships; Definition of containers; History of containers; Intermodal containers; Standardization of containers; Symbolism of containers
Le Container, 45–47
The Container (Der Behälter) publication, 38–40
Container bridge, 213–214
Container buildings and camps, 59–60, 287–292, 305–309, 333
Container City, 291
Container-contained theory, 67–68
Container format data, 238–241
"Container individual," 340–341
"Container order" of affluent societies, 304–305
Container Primer handbook, 50
Container principles, x–xi, 5–6, 35–36, 138, 301, 340
 in architecture, 64, 250
 criticism of, 138
 Matryoshka Principle, 68–69
 in packaging, 324–327
Container Security Initiative (CSI), vii–viii, 319–321
Container ships. *See also* Shipwrecked containers
 American Lancer, 49
 CMA CGM *Marco Polo,* 2
 Colombo Bay, 18–19
 Hansa Carrier, 29–30

Index

history of terminology, 42–44, 45, 50–51, 55–59, 84
Ideal X, 82–84, 115
introduction of, 82–85
invention of, 85–88
MSC *Napoli*, 1–6, 9, 27–29
MS *Fairland* (Sea-Land), 47–49
role of shipping lines, 192–197
Seatrain New York and *Havana*, 42–44
size, 2
"Container Ships," 44–45
Container shows (TV), 302–304
The Container Store, 335, 337, 338
Die Container Story documentary film, 49
Container supermarket, 330
Container Transport, 147
Controlled atmosphere, 228–229
Conveyer belts, 324
Corporate and social sculpture, 71
Cortada, James W., 220–221
Courier services. *See* Postal and courier services
Crespi, Silvio, 38–41, 119–120
Creveld, Martin van, 170
Crypt of Civilization, 14–15
CT-on-Line, 224
Culemeyer, Johann, 142
Curved landscapes, 76–79
Cybernetics
black box model, 216–219
self-controlling logistics, 237–238

Dachser company, 196
Data Communication System (DAKOSY), 224

Data processing
in art, 314–315
black box model, 68–69, 216–219
cloud computing, 240–247
container formats, 238–241
containers for, 228–230, 237–238
control and communication, 200–202, 221–227, 234
energy efficiency, 230–234
history of systematic research in development of container system, 208–216
and military logistics, 202–208
mobile data centers, 198–200, 202–203, 228–231, 241–247
programming concepts, 69–70
scope of, 230–234, 240–241
self-controlling, 235–238
software as boxes and containers, 69–70, 238–240
Data raft, 243–247
Davies, Colin, 284
De Cauter, Lieven, 337, 339
Definition of containers, 57–59. *See also* Symbolism of containers
in *Annals of the American Academy of Political and Social Science*, 45
in architectural context, 59–60
by *Bureau International des Containers*, 42–47
definition of boxes and chests, 147–148
in *Der Behälter* (The Container), 38–40

Definition of containers (cont.)
  in *Deutsches Wörterbuch* dictionary, 140
  in *Duden* dictionary, 138–139
  in *Great Complete Universal Encyclopedia of all Sciences and Arts*, 139
  history of terminology, 42–44, 45, 50–51, 55–59, 65–66, 84, 138–141
  in *Marine Engineering and Shipping Age*, 42–44
  by Meyercordt, 50–54
  in *Meyers Lexikon*, 141
  other words for, 60, 75
  uses for, 59–63
  vessels compared to containers, 138–140
DeLeu, Luc, 59
Deployment terminology, 202–203
Deportation and refugee camps, 301–309
Derrida, Jacques, 321
Dessau Bauhaus, 263, 271
Dessau-Törten residential development, 261, 283
Deutsche Bahn AG (German Federal Railway), 110, 121
Deutsche Post DHL, 172
The Digital Hand, 220
Diogenes, 130
Distribution centers
  regional, 157–161
  Wal-Mart, 154–157
Distriparks, 157–158
Dohrn, Reimer, 311
Donovan, Arthur, 106, 116
Doßmann, Axel, 62, 308

"A Dream Fulfilled: Living in Containers" news article, 289
Dressel 20 amphora, 134
Droysen, Johann Gustav, 17
Dr. Strangelove or: How I Learned to Stop Worrying and Love the Bomb film, 321

Easterling, Keller, 279–280
Ebbesmeyer, Curtis C., 30
Ebeling, Klaus, 181
Eckelmann, Kurt, 51
Economy. *See also* Logistics
  of containerized transport, 22–23, 36, 50, 62, 143, 146, 161–162
  global, 74
  images of, 75–81
  land- or sea-based, 95–98
  military influence on logistics, 202–208
  push-pull, 157
  and Wal-Mart Effect, 154–157
Econships, 113–115
Edison, Thomas, 242–243
"EDP System is Key to Door-to-Door-Moves," 224
Edwards, Paul, 337
Eickhoff, Hajo, 270–271
Einstein, Albert, 65
Empty-container logistics, 195
Energy efficiency of data processing, 230–234
Engels, Friedrich, 332
L'Esprit Nouveau (Toward an Architecture), 271–272
etoy, 70–72, 289, 314–315
Europallet, 188

Index

European and American transport, 49–55, 109–110, 186–188, 321
European Intermodal Association, 181
European Union (EU) refugee policy, 301–304, 310–312
Evergreen art project, 72
Evergreen Marine, 114, 194

Facebook, 242
Factories, 160
FedEx, 21
Felixstowe, England, 201
"Fish Story" photo and text project. *See* Sekula, Allan
Flow optimization, 169
Forbat, Fred, 259, 261
Fordism, 160, 162–163, 209
  in architecture, 249–250, 271, 274
Foreigners Out! *(Ausländer raus)*. *See* Schlingensief, Christoph
FOREST container (Verari), 230
Foucault, Michel, 336
Freedom Party of Austria, 302
"The Freight Container as a Contribution to Efficiency in Transportation," 45
Friedman, Thomas, 156, 334
Fröbel, Wilhelm August, 257
Froebel Gifts, 257, 258
From Bauhaus to Our House, 255
Fuel depots, 97
Full container load strategy (VL), 214–216
Funnel-Beaker Culture, 127
Fürth Repatriation Center, 307

Gans, Joshua S., 208
Gefäss (vessel), 139–140
Geographic positioning systems (GPS), 268
German Federal Railway *(Deutsche Bahn AG)*, 110, 121
German Institute for Standardization (DIN), 262, 269–270
Germany
  development of container transport, 41–42, 54, 98, 125
  pallets, 186–188
  portable buildings, 62
  railroads, 40–41
  rationalization, 163–164
  shipping preeminence, 195
Geyrhalter, Nikolaus, 325
Giedion, Siegfried, 323–324
Gilbreath, Frank B., 162
Globalization, 58, 74, 95, 108
  and Americanism, 55
  and brand consumption, 328–332
  computers as symbol of, 200, 217
  containers as symbol of, ix–x, 4–5, 72–79, 238, 291, 297–301
  and container transport, 84, 113–115, 126, 177
  and inequality, 305–315
  and native soil story, 90–92
  of supply chain, 23
  Tower of Babel metaphor, viii–ix
Globular Amphora Culture, 127
Goffman, Erving, 340
Gomorrah report, 313–314
Google, 242–244
GPS (geographic positioning systems), 268

Graham, Steven, 232
Grand Union Company, 327
Great Complete Universal Encyclopedia of all Sciences and Arts, 139
Great Western Railway, 92
Grimm brothers, 140
Gropius, Walter, 64, 259–262, 264, 271, 274, 275
GT Nexus, 225
Guantanamo prison, 306

Haider, Jörg, 302
Hamburg, Germany, 49, 125, 224
 stowaways, 310–312
Hannawald, Sven, 304
Hansa Carrier, 29–30
Hapag-Lloyd, 297
Harper's Weekly magazine (1873), 324
Heidegger, Martin, 75, 298
Heine, Heinrich, 92
Hellmuth, Yamasaki and Leinweber, 254
Hine, Thomas, 326–328
History of containers. *See also* Definition of containers
 amphorae and barrels, 129–138
 compared to boxes, 146–152
 concept of intermodality, 141–146
 early logistical research, 208–216
 introduction of computers, 220–227
 invention of intermodal containers, 85–90, 115–117, 203
 medieval, 131, 137, 141, 148–149
 mid-nineteenth century through 1960s, 118–121, 126, 222
 modern, 150–152
 MS *Fairland,* 47–49
 Neolithic and mythic, 127–129
 phases of, 125–127
 resistance to, 124–125, 192
 shipwrecked, 1–6, 9, 17–18, 24–36
Hitchcock, Henry-Russell, 275
Hoffmann-Axthelm, Dieter, 291
Holert, Tom, 308
Home and homelessness, 251–255, 278–280, 284–292, 333–334. *See also* Architecture
Honeycomb construction, 259–263
"Honeycomb Construction" exhibit, 259–261
Hoppe, Ralf, 239
Horizontal neighborhoods, 254
Hosoya, Hiromi, 328
Hughes, Thomas P., 274
Hugill, Peter, 98

ICE container (Rackable), 230
IKEA Distribution Center—Central Germany, 113
Iliad, 130
Illinois Institute of Chicago, 275
Immigration. *See* Refugees and asylum seekers
In-between connection, 176–177, 181, 184. *See also* Chains, logistical
Individual
 as container, 68, 335–337, 340–341
 development of, 298
 and social organization, 261

Index

Industrialization. *See* Rationalization; Standardization of containers
Inequality
  "container order" of affluent societies, 304–305, 311
  and piracy, 309–310
  and stowaways, 310–313
Infomediaries, 225
Ingraham, James, 30
Innis, Harold, 184
Intelligent packaging, 332–333
"Interchange and Container: The New Orgman," 281
Intermodal containers, ix–x
  classification system, 147–148
  and electronic data transmission, 223
  history of, 141–146
  invention of, 84–90
  logistics, 195–196
  postal and courier services, 174–175
  railway-ship, 24, 42–45, 92–93
  railway-truck trailer, 40–41, 44–45, 46
  Seaway, 42–45
  terminals and harbors, 112
International Congress of Modern Architecture (1928), 274
International Container Competition
  1930, 41
  1931, 146
International style of architecture, 273, 275–277
Internet, 225–227
  container format data, 238–240
  self-controlling logistics, 235–238
Interstate Commerce Commission, 90, 99, 106
In This World documentary fiction film, 312
INTRA, 225
ISO (International Standards Organization), 51–54
Isolation
  architectural, 253–255
  social, 336–337, 339
Isomorphies, 220

Jacobs, Thornwell, 14
Jarvis, William E., 12
Jencks, Charles, 251–252
Johnson, Mark, 66, 69
Johnson, Philip, 275
Jomini, Antoine-Henry, 169–170
Jumbo econships, 113–115
Jumping the Tracks: The Economic Battle between the Freeway and Railway—Container Transport, 142–143
Jünger, Ernst, 81

Kaurismäki, Aki, 333
Kazan, Elia, 75
Kittler, Friedrich, 185
Klein, Melanie, 68
Klein, Naomi, 328
Königsberg Bridge Problem (Euler), 159
Koselleck, Reinhart, 125, 165
Köstlin, Konrad, 17, 148, 149
Krusemark, Max, 142–145
Kubrick, Stanley, 321
Kurakawa, Kisho, 283

Lakoff, George, 66, 69
Land, conception of
  images associated with containers, 102–103
  Melville, 101–102
  Nietzsche, 100–101
  terminals and harbors, 112
Land and Sea. *See* Schmitt, Carl
Land bridges for containers, 103–112
  American compared to European, 109–110
  stacktrains, 107–110
Land-water system. *See* Sea-land transport
Läpple, Dieter, 170
Larkin building (Frank Lloyd Wright), 256
Lawrence, T. E. (Lawrence of Arabia), 323
Laws
  Container Security Initiative (CSI), vii–viii
  international regulations, 28
  Public Law 110–53, vii, 319
  regulation of container transport, 41, 90, 106, 109, 115
  Schengen Agreement, 310
  Staggers Act (1980), 106
Le Corbusier (Charles-Edouard Jeanneret), 64, 248–250, 263–264, 266, 271–272, 289
Lego metaphor, 75
Leo VI (Emperor), 169
Less than container load (LVL), 214–216
Levinson, Marc, 23, 55, 116
Liftvans, 118–120, 122
Lignano, Giuseppe, 287

"Little Boxes" song, 278–279
Living machine (Le Corbusier), 250, 254
Logistics. *See also* Container principles
  as a branch of economics, 168–172
  of data processing, 200–208
  definitions, 162, 168–171
  early systematic research, 208–216
  history of, 168–170
  and media, 184–186
  of military supply, 169–171, 189–191, 202–208
  of modernity, 10
  organization of chains, 176–180
  pallets, 186–188
  parasite model, 181–184
  practical inefficiencies, 189–192
  and rationalization, 161–167
  self-controlling, 235–238
  of suburban housing, 280–282
  third-party impact, 192–197
  Wal-Mart Effect, 154–157
Logistics centers. *See* Distribution centers
LogMotionLab, 236
London, England
  architecture, 289, 291
  Great Western Railway, 92
  and international regulations, 28
  port, 300–301
Looting of containers from MSC *Napoli*, 2–4, 9
Lot-Ek, 287–290
Luhmann, Niklas, 316

Index

Maersk, 114, 192–193
Mafia, Neapolitan, 313–314
Magnified System Diagram, 2000
 drawings, 296
Mahan, Alfred Thayer, 95
La Maison Standardisée, 249
Mann, Heinrich, 322
Manufacturing, 160
The Man without a Past (film by Aki Kaurismäki), 72, 333
Marine Engineering and Shipping Age publication, 42–44
Marx, Karl, 329, 331–332
Mass production, 160
Matryoshka Principle and functions, 68–69
Matson Navigation Company, 208–214
Matthew Marks gallery, 296
McLean, Malcom, 24, 84–92, 107
 economic efficiency, 86–88, 99
 invention of container ships, 85–90, 115–117
 jumbo econships, 113–115
 native soil story, 90–92
McLean Trucking Corporation, 86–90
McLuhan, Marshall, 184
Mean Time, 2000 (exhibition piece by Darren Almond), 295–297
Mechanization Takes Command, 323–324
Media theory, 184–186
Mediterranean Shipping Company (MSC), 28, 193–194
Melville, Herman, 101–102
Metacontainer, 19, 74, 184, 218, 316, 329
Meyer, Adolf, 259, 261
Meyercordt, Walter, 50

Microcontainers, 339
Mies van der Rohe, Ludwig, 263–264, 275
Mies box, 276–278
Military domination, 93–97
Military supply logistics, 169–171, 189–191
 influence of operations research, 202–208
Mission Eternity, 314–315
Mobile data centers
 Project Blackbox, 199–200, 202, 228–231, 240, 242
 water-based, 241–247
Mobile dwelling unit (MDU), 287–288, 290
Mobilization, 95
Modified atmosphere, 229
Modularization, 5, 60, 334–336, 340. *See also* Rationalization; Standardization of containers
 of architecture, 248, 250, 261–262, 265, 271, 275, 282–284
 of computers, 199–202, 242
 of education, 335
 and logistics, 160, 162, 165
The Mole in the City cartoon film, 281–282
Møller, A. P., 114, 192
Møller, Peter Maersk, 192–193
Monsun yacht collision, 25
Morgenstern, Oskar, 205
MSC Napoli container ship, *xvi*, 3
 accident, 1–6, 9
 history of, 27–29
Müller, Felix, 182
Muller, Gerhardt, 190
Mumford, Lewis, 252, 340
Murschetz, Luis, 311

Nagakin Capsule Tower, 283
Napoleon, 322
Napolitano, Janet, 321
Napster, 245
Nation-states as containers, 337–339
Native soil story, 90–92
Nazi era, 150
  architecture, 264–265, 275
Neolithic Age, 127–128
Nerdinger, Winfried, 274
Nesting containers, 68–69
Neubert, Christoph, 237
Neufert, Ernst, 264–270
Neurath, Otto, 79, 80
New York City
  On the Waterfront film, 75, 87
  Port Elizabeth, New Jersey, 222
  port facilities, 124–125
Nietzsche, Friedrich, 66, 100–101
9/11 reaction, vii–viii, 306, 313, 318–322
1984 metaphor, 252
No Logo: Taking Aim at the Brand Bullies, 328
Nonvessel operators (NVOs)
  Mediterranean Shipping Company (MSC), 28, 193–194
  postal and courier services, 172–174
"Notes on the Formulation of the Theory of Logistics," 205

Obama, Barack, 306
Obermaier, Robert, 182
Ocean current research, 29–33
Olive oil transport, 134–135
On the Waterfront film, 75, 87
Ontological metaphor, 66

Operations research, 208–216
Operations Research journal, 213
Optimization
  as basis for modern logistics, 160, 168–169
  and computers, 221–222
  flow, 169
Oracle, 229–230
The Organization Man, 279
Orsenna, Érik, 32
Orwell, George, 252
Ostwald, Wilhelm, 262, 268
Our Daily Bread documentary film, 325

Pachnicke, Peter, 129
Pacific Coast Engineering Company (Paceco), 214
Packaging, industrial, 323–333
  adaptation to, 325–326
  and brands, 328–332
  as communication medium, 326–327, 332–333
  disposal of, 331–332
  revolutionary technology, 323, 326, 329
  social function, 326–328
Packet boats, 175–176
Pagonis, William, 207
Pallet Pool, 188
Pallets, 186–188
Panama Canal, 2, 103, 114, 309
Pan Atlantic Steamship Company, 83–84, 89, 99
Pandora's box metaphor, 130, 317–318
Parasite model of logistics, 181–184
Pendray, G. Edward, 12, 23, 27

Perret, Auguste and Gustave, 249
Peukert, Detlev, 164
Physics-based models, 64–66
Pias, Claus, 151
Pictorial representation, 76–81
Piggyback model, 182
Pirate Bay, 245
Pircher, Wolfgang, 203
Pithoi (vessels), 129–130, 317. *See also* Amphorae
Platform principle, 147–148
Playtime film, 277–278
PMDC container (IBM), 230
POD container (Hewlett-Packard), 230
Poet, Paul, 304
Politics in Independent Theater Festival *(Festival Politik im Freien Theater)*, 72–73
Pollak, Richard, 18–19
Pompeii as time capsule, 16–17
Pörksen, Uwe, 74
Porstmann, Walter, 262, 268
Portable containers, 61–62
Port Elizabeth, New Jersey, 222
Postal and courier services, 172–176
  Deutsche Post DHL, 172
  history of, 172–174
  intermodality, 174–175
  as nonvessel operators (NVOs), 172–174
  and packet boats, 175–176
  as paradigm, 172
Postcards, 175
Post-Panamax container ship, 2
Powell Jr., Stanley, 209
The Power of Identity: The Information Age, 297

Power-Stow, 226
Précis de l'art de la guerre (The Art of War), 169–170
Prigge, Walter, 263
Privacy, 305
Program for International Student Assessment Studies, 335
Pruitt-Igoe project, 251–255, 254
Psychoanalytic theory, 67–68
Public Law 110–53 (United States), vii, 319
Push-pull economic models, 157

Quarterly Review magazine (1839), 93

Railway transport
  American compared to European, 109–110
  American President Lines (APL), 106–107
  Grand Union Company, 327
  history of, 40–41, 120, 141–147, 150
  as medium, 185
  Seatrain, 42–45
  stacktrains, 107–110
RAND (Research and Development), 204
Rand Corporation, 204–205, 320
Rathenau, Walther, 261
Rationalization, 36, 161–167
  in architecture, 249–252, 259–282
  definition of, 163–164
  economic impact of, 166–167
  ideology of, 164–165
"Rationalization" article, 165

Rationeller Transport magazine, 165, 178, 186
Refrigerated containers (reefers), 228–229
Refugees and asylum seekers, 301–309
 Chinese, 313–314
 Nigerian, 311–312
 Romanian, 312–313
 stowaways, 310–313
Regulation of container transport, 41, 90, 106, 109, 115
 international, 28
Relay vehicle, 142–145
Research Association for Container Transport (1928), 146
Reuse of containers, 341
 architectural, 287–292
Revelation of John metaphor, 101–103, 108–109
Reynolds, Malvina. *See* "Little Boxes"
RFID technologies, 235–238
Rheinberger, Hans-Jörg, 33
Robie House (Frank Lloyd Wright), 258
Roman Empire, 130–131, 134
RoRo ("Roll on, Roll off") ships, 116, 189
Rotterdam, Netherlands, 47, 125, 224
RR documentary film, 107–108

Sarcophagus Tank multimedia piece, 314–315
Saviano, Roberto. *See Gomorrah* report
Schaefer, Markus, 328
Scharmen, Fred, 157–158

Schengen Agreement, 310
Schlingensief, Christoph, 301–304
Schmitt, Carl, 93–95
Scientific management, 162
Sea, conception of
 American influence, 98–100
 impact of steamships, 90–93
 territorial-historical, 93–98
Sea-Land container ships
 Ideal X, 82–84
 MS *Fairland*, 47–49
Sealand micronation, 245
Sea-Land Services, Inc.
 acquisition by Maersk, 193
 advertisement, 96
 establishment of company, 84–90, 124–125
 financing for, 89–90
 first U.S.-European container transport service, 47–49
 first voyage, 82–84
 introduction of computers, 222
 as military supplier, 190
 name of, 117
Sea-land transport
 globalization, 113–115
 impact on conception of land, 100–112
 impact on conception of sea, 90–100
 invention of, 85–90, 115–117
Sears, Roebuck and Company logistics, 207
Seaton, Bruce, 106
Seatrain, 42–45
 advertisement, 104–105
 diagram, 43
 minibridge service, 104

Index

Security measures, vii–viii, 306, 313, 318–322
Seeger, Pete. *See* "Little Boxes" song
Sekula, Allan, 334
Self-controlling logistics, 235–238
September 11, 2001. *See* 9/11 reaction
Sequential disintegration/constructivism, 22–24
Serres, Michel, 46–47, 181–183
Servan-Schreiber, Jean-Jacques. *See American Challenge*
Shipwrecked containers, 17–18, 24–36
  Hansa Carrier, 29–30
  magnitude and costs of, 24–27, 31
  MSC *Napoli*, 1–6, 9, 27–29
  and ocean current research, 29–33
Shoe spill, 29, 31, 33
Siegert, Bernhard, 5, 93, 175, 226–227
Silk Road, industrialized, 109
Silos, 297
Size of container ships
  CGM *Napoli*, 2
  CMA CGM *Marco Polo*, 2
  Triple E Class, 2
Sloterdijk, Peter, 62, 74, 112, 262–263, 279, 315, 322, 333
Snow Crash, 245–247
Social-scientific spatial models, 64–66
Space-time regimen of containers
  in art, 294–297, 301–304, 314–315, 329
  global advertising claims, 297–299
  immortality and eternity, 313–315
  impact of local conditions, 300
  symbolism of, 19–20, 73, 295–298, 334–335
Spain, 308
Spatial models, 64–66
Specialization, 143, 160–161
Spheres trilogy, 262
Stacktrains, 107–110
Staggers Act (1980), 106
"Standardization and Housing Shortage," 259
Standardization of containers, x–xi. *See also* Modularization; Rationalization
  architectural, 249–252, 259–282
  Bureau International des Containers, 45–47
  early logistics research, 208–216
  early twentieth-century, 122
  history of, 137–138, 162
  industrial packaging, 323–333
  ISO, 51–54, 188
  lack of, 189
  as links in chains, 335
  as metaphor, 76, 81
  pallets, 188
  postal service, 175
  and software communication, 200
The Standardized House, 249
Steamship, impact of, 92–99
Stephenson, Neal. *See Snow Crash*
St. Louis, Missouri, 254
Stowaways, 310–313
Stowaways working group, 311

Studer, Manfred, 311
Subjectivity, 332, 335–341
Suez Canal, 103, 114
Sun Microsystems Project Blackbox, 198–200, 199–200, 202, 228, 229, 240, 242
Sun Modular Datacenter S20 (MD S20), 228–231
Supply chain management, 178–180
Swarttouw, Frans, 125
Symbolism of containers. *See also* Definition of containers
 Americanism, 54–55
 apocalypse metaphor, 101–103, 108–109
 in architecture, 59–64, 250, 251, 255, 278–282, 283, 287
 in art, 66, 70–74, 259–261, 278–279, 295–297, 333
 bee metaphor, 261
 as cornucopia, 130
 in cultural context, 25–26, 29–30, 33–35, 66, 127–129
 in data processing, 68–72, 185–186, 200–201, 238
 in economic context, 50, 76–81, 85–90, 92–100
 in emigration history, 9–10
 fear of globalized terrorism, 313, 317–322
 as icons of globalization, ix–x, 72–76
 in international politics, 93–95, 337–339
 Lego metaphor, 75
 as links in chains, 178–180, 335
 logistics and rationality, 36, 57–58, 69, 75–76
 in media, 302–304
 as medium of transport, 46–47, 184–186
 native soil success story, 90–92
 nesting containers, 68–69
 1984 metaphor, 252
 ontological metaphor, 66–67, 69–70
 Pandora's box metaphor, 130, 317–318
 as parasite, 181–184
 and personal identity, 92–100, 335–337
 psychotherapist as, 67–68
 religious and mythical, 73–74, 128–129, 140
 Revelation of John metaphor, 101–103, 108–109
 standardization metaphor, 76, 81
 in time-space relationship, 19–20, 64–66, 73, 226–227, 295–298
 Tower of Babel metaphor, viii–ix

Taggart, Stewart, 220
TANKS (etoy), 71–72
Tanner, Jakob, 77
Tapping in the Container TV show, 310
Tati, Jacques, 277–278
Taut, Bruno, 256–257
Taylor, Frederick W., 162
Taylorism, 163, 251, 264, 273–275
Temporary containers, 61–62
Terkessidis, Mark, 308
Terra Firma Capital Partners, 102
Third-party logistics, 192–197
 boxes and containers, 195–196
 infomediaries, 225
 transport industry, 221

Index

Third-party shipping lines, 192–195
  Evergreen Marine, 114, 194
  Maersk, 114, 192–193
  Mediterranean Shipping Company (MSC), 28
  tramp shipping companies, 194–195
3PL. *See* Third-party logistics
Time. *See* Space-time regimen of containers
Time capsules, 9–38
  Andy Warhol, 10–12
  containers as, 19, 23–24
  Crypt of Civilization, 14–15
  1876 World's Fair, 12
  history of terminology, 12, 23, 27
  interrupted shipments, 20–22
  Pompeii, 16–17
  unplanned, 15–16
  Westinghouse Time Capsule of Cupaloy, 12–14
Tokyo, Japan, 283
Tolla, Ada, 287
The Total Package: The Secret History and Hidden Meanings of Boxes, Bottles, Cans, and Other Persuasive Containers, 326–328
Toward an Architecture (L'Esprit Nouveau), 271–272
Tower of Babel metaphor, viii–ix
"Trailer Park Computing" blog commentary, 242–243
Tramp shipping companies, 194–195
"Transport Chains, Integrated Transit and Containers—Key Points in Business Logistics," 178
Transport Medium exhibit, 296
Traveling Salesman Problem, 160
Triple E Class container ships, 2, 114
Truck intermodal containers, 40–41, 44–45, 85–90
"The 20-Ton Packet: Ocean Shipping Is the Biggest Real-Time Data-Streaming Network in the World," 220

United Food and Commercial Workers, 318
Unité d'habitation (unit of habitation), 271–273, 289
United Kingdom, 94–95, 97–98
United States-Mexican border, 309
United States transport. *See* American transport
Urban Space Management, 291
Urnfield Culture, 127
U.S. Lines, 113, 117
Utriclarii, 135–136

Value, economic
  conceptual value added, 328
  use value and exchange value, 329–331
Vessel (*gefäss*), 139–140
Vienna Festival (2000), 301–302
Vietnam supply system, 190, 222
Virilio, Paul, 27
Virtualization. *See* Cloud computing
Visiotype, 74
Von Neumann, John, 205

Wal-Mart
  container security, 318
  distribution center, 154–157
  logistics, 208
  package processing, 158
Warhol, Andy, 329
  Museum, 11
  Time Capsules, 10–12
Weldon, Foster S., 208–214
Welke, Ulrich, 191
Wenzel, Jan and Kai, 62, 308
Westinghouse Time Capsule of
    Cupaloy, 12–14
"What Is a Container?," 38–39
Whore of Babylon metaphor,
    101–103, 108–109
Whyte, William H., 279
Wine transport, 135–136
Winterbottom, Michael, 312
Wolfe, Tom, 255, 275, 276–277
World Automobile Congress
    (1928), 41, 146
The World Is Flat, 156, 334
World of Vessels—from Antiquity
    to Picasso exhibit, 129
World picture, 75–76
World's Fair (1876), 12
Wright, Frank Lloyd, 255–258
Wriston, Walter, 99–100

Yale box, 276–278

Zedler, Johann Heinrich, 139
Zim Israel Navigation Company,
    287